Born and raised in Los Angeles, ...hen
received his B.A. in history...
Cali... at Berkel...
Co...

D1551676

ANATOMY and DESTINY

ANATOMY and DESTINY

A Cultural History of the Human Body

by STEPHEN KERN

THE BOBBS-MERRILL COMPANY, INC.
INDIANAPOLIS · NEW YORK

Chapter One, "The Onset of Victorian Sexual Morality," has appeared in modified form as "The Onset of Bourgeois Sexual Morality" in *Book Forum*, Fall 1974.

Chapter Five, "Olfactory Ontology and Scented Harmonies," has appeared in *The Journal of Popular Culture*, Spring 1974.

Portions of Chapter Ten, "Physiology of the Victorian Family," have appeared as "Explosive Intimacy: Psychodynamics of the Victorian Family" in *The History of Childhood Quarterly*, Summer 1974.

For permission to use excerpts from copyrighted material, grateful acknowledgment is made to the following:

Grove Press, Inc., publishers of TROPIC OF CANCER by Henry Miller, copyright © 1961 by Grove Press, Inc.; Pantheon Books, Inc., a division of Random House, Inc., THE SLEEPWALKERS by Herman Broch, translated by Edwin and Willa Muir, copyright © 1947 by Pantheon Books, Inc.; New Directions Publishing Company, publishers of NAUSEA by Jean-Paul Sartre, translated by Lloyd Alexander, copyright © 1964 by New Directions Publishing Corporation; Philosophical Library, Inc., publishers of BEING AND NOTHINGNESS by Jean-Paul Sartre, translated by Hazel E. Barnes, copyright © 1957 by Philosophical Library, Inc.; McGraw-Hill Book Company, publishers of KARL MARX: *Early Writings*, translated and edited by T. B. Bottomore, copyright © 1963 by T. B. Bottomore; W. W. Norton & Company, publishers of THE MARX-ENGELS READER, edited by Robert C. Tucker, copyright © 1972 by W. W. Norton & Company, Inc.

The publishers wish to thank the following for photographs and plates which appear in this book:

Caisse Nationale des Monuments Historiques, Paris, for plates 4, 6, and 10; Clichés Agraci, Paris, for plates 7 and 11; Lichtbildwerkstätte 'Alpenland,' Vienna, for plate 17; J. E. Bulloz, Paris, for plate 14; S. P. A. D. E. M., Paris, for plate 12; Photographie Giraudon, Paris, for plates 3 and 5; Cliché des Musées Nationaux, Paris, for plates 8 and 13; Photo Meyer K. G., Vienna, for plate 18.

Copyright © 1975 by Stephen Kern

All rights reserved, including the right of reproduction
in whole or in part in any form
Published by the Bobbs-Merrill Company, Inc.
Indianapolis   New York

ISBN 0-672-52091-5

Library of Congress catalog card number 74-17651
Designed by Helga Maass
Manufactured in the United States of America

First printing

To my parents

Table of Contents

INTRODUCTION		ix
Chapter 1:	THE ONSET OF VICTORIAN SEXUAL MORALITY	1
Chapter 2:	THE SARTORIAL ASSAULT ON THE BODY	10
Chapter 3:	THE BODY IN ART	21
Chapter 4:	CONTAGION AND HUMAN RELATIONS	34
Chapter 5:	OLFACTORY ONTOLOGY AND SCENTED HARMONIES	45
Chapter 6:	SEXUAL ALIENATION AND SEXUAL SELECTION	56
Chapter 7:	MATERIALISM AND THE MIND-BODY PROBLEM	67
Chapter 8:	THE BODY ELECTRIC	80
Chapter 9:	PURE WOMEN AND SUPERB MEN	95
Chapter 10:	PHYSIOLOGY OF THE VICTORIAN FAMILY	109
Chapter 11:	THE SCIENTIFIC STUDY OF SEX	125
Chapter 12:	SEXUAL PATHOLOGY AND HOMOSEXUALITY	139
Chapter 13:	WHAT A YOUNG GIRL WOULD KNOW WHO KNEW *WHAT A YOUNG GIRL OUGHT TO KNOW*	153

Chapter 14:	THE DISCOVERY OF THE MEANING OF THE BODY	169
Chapter 15:	EROS IN BARBED WIRE	191
Chapter 16:	CRAZY COCK AND LADY JANE	207
Chapter 17:	BODY POLITICS IN GERMANY	221
Chapter 18:	THE PHILOSOPHY OF THE BODY	238
CONCLUSION		251
AUTHOR'S NOTE		258
NOTES		260
INDEX		293

INTRODUCTION

Ours is an age obsessed with youth, health, and physical beauty. Television and motion pictures, the dominant visual media, churn out persistent reminders that the lithe and graceful body, the dimpled smile set in an attractive face, are the keys to happiness, perhaps even its essence. This celebration of bodily excellence is in some respects a welcome repudiation of the sexual repression that has dominated much of our past, but it does not solve entirely the great problem of human corporeality that has troubled mankind throughout history. The twentieth century may have learned to shorten bathing suits and get the body out into the light of day; it may have lightened clothing to allow freer movement and fuller display of the beguiling contours underneath; and it may have campaigned for sexual liberation. But it still has not fully come to terms with the human body.

The body is a source of great anxiety, derived above all from the fear of death. When this ingenious structure of protoplasm breaks down, life stops, and all our striving and learning come to an abrupt end. The body is also a source of pain. It is easily crippled and inevitably succumbs to illness and deterioration. However "young at heart" we try to remain, the passage of time inevitably marks our bodies. Wrinkles, sagging flesh, unpleasant odors relentlessly remind us of the decline of our bodies, while our mental capacities continue to flourish.

Even the pleasures have troubled mankind. Over the ages moralists have devised various arguments about the unimportance of our corporeal existence as much to belittle the pleasures of life as to steel man against pain. The body also has needs that require constant attention. Most of our energy goes to feeding, clothing, and caring for our bodies, and the sexual instinct requires still more energy.

Perhaps most disturbing is the way the body intrudes itself into the heart of our existence. Man takes pride in his personality—the mind, soul, ego, or whatever term he uses for that central core he believes to be his true self—and is insulted by the way his body seems to dominate it. A man who is five feet tall is free to determine how he will live, but he can never escape the confines of his physical stature. He is condemned to live his five feet, and no matter how big his soul, he will have to look up at most people his entire life. He is compelled to compensate for an aspect of his existence he can never control. If a person is strikingly fat or thin, healthy or sick, beautiful or ugly, his existence is profoundly affected. We are locked in our bodies, and no amount of study or integrity or genius can alter them. There is perhaps some consolation in the comment of Camus that we are all responsible for the way we look after forty. But during the most important years of adolescence and early adulthood the gift of good looks or the curse of ungainliness can dominate entirely our image of ourselves.

The body also influences our social existence and our relations with other people. A beautiful woman generally leads a vastly different life from a plain one. The beautiful child often receives more attention and affection than does the less beautiful one. Obese children are consistently teased by their peers and stereotyped as lazy and apathetic by adults. The rejection experienced by an adolescent because he is homely can be as significant in shaping his personality as the effects of social, or even racial, prejudice—racial prejudice also separates people by physical attributes. Aristocratic, bourgeois, and proletarian sexual aesthetics share a common appreciation of physical beauty.

Children see this problem in the story of the Ugly Duckling, who is cruelly abused for being ugly. The implied moral is that if you are so abused as a child, fear not, because you will become beautiful when you grow up—hardly a comfort to those who suspect that they may not. When the Ugly Duckling finally discovers that he has become a beautiful swan, he pauses to remind himself that "a good heart never becomes proud," but the story ends with his ecstasy over his newly discovered beauty: "I never dreamed of so much happiness when I was the Ugly Duckling!" Of course, moralists warn that "beauty is only skin deep" and that "handsome is as handsome does," but such moralizings often go unheeded when people pair off sexually. Sexual needs are a source of great conflict. Man craves tenderness, but it is often forgotten during the sexual act when the desire of the moment obliterates the gentler emotions. He seeks understanding but may find that he is sexually attracted to someone who is not. It is the height of absurdity that a couple, suited to each other in every other respect, may be sexually incompatible.

For a number of reasons, therefore, mankind finds its corporeal nature a problem and has devised many ways of dealing with it. This book is a history of the cultural record left since the mid-nineteenth century as Europeans and Americans began to understand and accept their corporeality. Chapter One outlines some of the more striking examples of the restrictive sexual morality that dominated Western culture in the early Victorian era. The material in the remaining chapters is arranged topically and chronologically to illustrate the gradual dissolution of that morality. Chapters Two and Three survey developments in fashion and art throughout the nineteenth century, which show a growing willingness to liberate the body from constricting apparel and from rigid artistic conventions. Chapters Four and Five discuss progress in public hygiene and its effect on human relations. Chapters Six through Eight concern some of the leading intellectuals of the age whose works explore the corporeal determinants of life and the relation of mind and body. Chapters Nine and Ten recount theories on the bodily determinants of gender roles and family relations. With Chapter Eleven the text follows a more chronological course, beginning with a discussion of scientific studies of sexuality and the literature on sex education that was available around the turn of the century. Chapter Fourteen evaluates Freud's contribution to the mind-body problem, and Chapter Fifteen surveys the impact of the First World War on attitudes toward the body. The final three chapters provide a look at some important developments during the postwar years in England, America, Germany, and France: the fiction of D. H. Lawrence and Henry Miller, the Nazi physical culture movement, and Sartre's philosophy of the body.

Although the presentation of the material is largely descriptive, there is a central argument that since the heyday of Victorianism progress has been made toward better understanding of the body and a freer indulgence in bodily pleasure. Not all of the material, however, illustrates uninterrupted progress. When evidence ran counter to the general argument, I sacrificed consistency for the complexity of the fuller story. There were opponents to clothing reform, sexual reform, and the "emancipation of the flesh" throughout the period, and I have introduced their ideas when they help illustrate the climate of opinion that generated the calls for greater liberation. Thus, James Buchanan's bitter opposition to the "fleshly" poets of his age gives a fuller picture of the shocking impact of Whitman, Rossetti, and Baudelaire. A good deal of the literature on sex education shows the persistence of Victorian sexual morality, and I have presented that material to illustrate more sharply the enormous impact of World War I. The Nazi physical culture movement was a mixture of progressive and reactionary ideas that illustrate both the movement to cultivate physical excellence on a national scale and the racism to which such a national endeavor can lead. I hope the reader will abide the occasional digressions and summaries of counter-evidence in the interest of a broader understanding of the variety of ideas that have been recorded.

I do not mean to suggest with this discussion of "progress" that the problems of human corporeality have been entirely solved. The anxieties summarized in the opening pages of this introduction are largely universal. One age may learn to cope with them more successfully than another, but the problems remain. Nevertheless, evidence strongly suggests that over the past century and a quarter Europeans and Americans have acquired a great deal of knowledge about the human body and have begun to put that knowledge to use. Although progress since the mid-nineteenth century was slow and often interrupted by denial and repression, the broad picture shows increasing efforts to enjoy the human body and to understand the manifold ways it influences life.

Chapter 1

The Onset of Victorian Sexual Morality

In the early decades of the nineteenth century European sexual morality shifted decisively from the permissiveness of the previous century to the severely restrictive rules and attitudes that have come to be characterized as Victorianism.[1] Sir Walter Scott noted this change when his great-aunt, whom he had supplied with the works of a seventeenth-century novelist, Aphra Behn, returned them with the suggestion that he throw them into the fire. "I found it impossible to get through the very first of the novels," she remarked. "But is it not very odd that I, an old woman of eighty and upwards, sitting alone, feel myself ashamed to read a book which sixty years ago I could have heard read aloud for large circles consisting of the first and most credible society in London?"[2] The shame which Scott's aunt felt was the result of an increasing denial and repression of sexuality which reached a high point during the middle decades of the century. I should like to survey some of the ideas and practices that characterized that repression.

Those Victorians who dominated public morality came to regard their bodies as a threat to respectability, and their attitude toward them was a combination of denial, distortion, and fear. Social historians have reported patients, particularly female patients, refusing to submit to medical examinations by doctors. The first female physician in America, Elizabeth Blackwell, explained that she was partly motivated to enter medicine by the death of a woman friend of hers who had refused treatment by male physicians.[3] The physician who developed the diaphragm in Germany reported several instances of extreme prudery among his female patients. The mother of one patient suffering from inflamed ovaries refused to let him give her daughter a necessary rectal examination and protested angrily that she "had never heard of anything so vulgar."[4] Some doctors resorted to using dummies on which the patients showed where they felt pain. "The doctor was then allowed to touch the place through the under clothing, or through a cloth—even that, of course, only in the presence of the patient's husband or mother."[5]

The most widely recounted instances of excessive prudery involved the concealment of women's legs. Books on etiquette repeatedly advised their readers to avoid mention of "legs" altogether or, when this proved unfeasible, to use the less suggestive word "limb." One Englishman traveling in the United States in the 1830s reported having seen trousers covering the legs of a pianoforte in a seminary for women. The headmistress "had dressed all these four limbs in modest little trousers, with frills at the bottom of them!"[6] Freud's son Martin reported that though he lived with his Aunt Minna for many years, he "never had any realization that she had legs."[7] The campaign to conceal the leg was so effective that by mid-century men were easily aroused by a glimpse of a woman's ankle. The high incidence at this time of fetishes involving shoes and stockings further testifies to the exaggerated eroticism generated by hiding the lower half of the female body.

Not just the legs but the entire body was covered with a great deal of clothing. After a short flirtation with lightweight and scant attire during the Directory and the Napoleonic periods in France, women's clothing around 1820 began to cover the entire body with scrupulous thoroughness. The hemline dropped to the ground and remained there for a century. Long sleeves were extended by lace, high collars covered the neck (with the single exception of the high fashion décolleté style for evening attire), while gloves and bonnets completed the mummification. It was not enough just to cover the female body; Victorian clothing also abused it. Shoes disfigured the feet and garters marred the legs, but the major assault on the body was achieved by the corset, which constricted breathing, circulation, digestion, and movement. Any inclination women might have had to exercise or enjoy their bodies was impeded by the corset.

Attitudes regarding sexuality ranged from moral condemnation to disutility. Sexual excitation was regarded as sinful, and masturbation was cited as a cause of a host of diseases ranging from warts and anemia to sexual pathology and insanity. Nocturnal emissions (pollutions) were believed to lead to chronic involuntary discharge of semen—a disease called "spermatorrhea." Sexual activity was regarded generally as a waste of energy except in the service of procreation, and excessive indulgence was believed to lead to general physical exhaustion. Some marriage manuals argued that more than one orgasm a week might debilitate a man and take him away from more productive labors. Religious and medical tracts plied young men about the evils of their sexual feelings, and sexuality came to be regarded as sinful even when pursued under the blessings of religious sanction. Pleasure itself came to be suspected as self-indulgent, sinful, and wasteful. Venereal disease was a constant source of anxiety, particularly to the young men who satisfied their sexual needs with prostitutes, and the treatment for venereal disease, often administered by quacks, offered no certain cure and was painful and dangerous. And in an age when the use of contraceptives was restricted by limited availability and faulty sex education, the danger of pregnancy was ever present for those women who did challenge the going sex ethic.

In this atmosphere of ignorance and fear, male-female relations were strained and anxiety-ridden. An often-repeated bit of gossip on the nature of marital intimacy of the age concerned the marriage of Robert and Elizabeth Browning. These two poets, who gave England some of its most lyrical love poems and became a model of marital happiness, are said to have never seen each other naked during a lifetime of marriage.

Women had a particularly conflict-ridden sexual role to play. On the one hand, they were supposed to be passive during the sex act and were to show no pleasure. Marriage manuals before the 1870s did not generally discuss female orgasms.[8] But in contradiction to this denial of female sexuality, women were led to believe that in other respects they were totally physical beings, instinctive and emotional by nature, constructed essentially for child-bearing and nursing. Women reading these manuals were torn with contradictory ideas about the extent to which sexual feelings were to determine their behavior.

Men, on the other hand, were showered with advice about manliness, involving hardness and aggressiveness. A double standard dominated the rules of extra-marital relations so that men could seek sexual satisfaction outside the marriage while women could not. Courtships were periods of prolonged frustration, and the wedding night often culminated in a rape. The conspiracy of silence about sexuality frequently let young girls enter puberty ignorant of menstruation. Their horror was matched by the young man's fear that his first ejaculation was a sign of some disease. Many women entered marriage without any understanding of what their marital sexual role would be. They had perhaps never been alone with their husband before the wedding night and may never have touched any part of him. He, on the other hand, had most likely been introduced to sex by prostitutes and had been forced to contain his sexual desire for his fiancée until the wedding night. The going idea of manliness involved being forceful and masterful, and therefore he was not likely to guide his innocent new bride with the patience she needed. It is not surprising that a number of manuals toward the end of the century began to caution young husbands against ruining their marriages on the first night.

Such was the sexual morality that dominated the lives of the upper classes of Europe in the mid-nineteenth century. The remaining chapters will trace the gradual dissolution of this sexual morality in the late nineteenth and early twentieth centuries, but to complete my introduction to the moral climate of the period, I must first discuss some of the theories that have been suggested by historians of the early nineteenth century to explain why this particular sexual morality became so influential at that time.

One line of reasoning has focused on the immediate impact of the collapse of the French monarchy in 1789, which generated a re-evaluation of morality among the upper classes all over Europe. A prevailing concern was that the gambling, drinking, partying, dueling, wenching, and general moral laxity that, at least in retrospect, had seemed to permeate the aristocracy in France had contributed to its downfall. In England, criticism of the excesses of the terror in France was coupled with demands for reform of morals to fortify the country against the dangers of Jacobinism. The prevailing mood among the ruling classes was expressed in a pamphlet by John Bowdler (brother of the famous expurgator, Thomas Bowdler) entitled "Reform or Ruin," which argued that the English ruling classes would also be swept from power if they did not reform their life style. Another zealous reformer of the time, William Wilberforce—head of the Society for the Suppression of Vice, founded in 1802—saw an "intimate connection between moral levity and political sedition among the lower classes." [9]

Some historians have concluded that the sexual morality of the early nineteenth century can be explained as a direct result of the needs of the bourgeoisie, which had recently come to power. The emergent bourgeoisie was as anxious to protect its new power as the aristocracy was to avoid revolution and ruin. The middle classes were intent on working out a life style that would avoid the moral corruption that they believed had been weakening the upper classes. And so in the early decades of the century there appeared a number of books on etiquette which worked out the intricacies of bourgeois propriety. They also wanted to dissociate themselves from the violence and the sexual promiscuity associated with the lower classes, and consequently their sexual morality began to become ever more restrictive. Harold Nicolson suggests that at this time the upper ranks of society were "suddenly invaded by the newly rich: the snobbishness of the period was predominantly a defense apparatus." [10] Wilhelm Reich offered a bold speculative interpretation of the effect of the bourgeoisie upon European sexual morality. The revolutionary bourgeoisie of the eighteenth century campaigned against the sexual morality of the Church, he noted, but as it became established as a ruling class it allied itself with the Church to help control the proletariat that it found so threatening. The bourgeois family suited the requirements of the capitalists insofar as it supplied an institution to enlarge and maintain property holdings and facilitate the transmission of inheritance. According to Reich, the bourgeoisie insisted upon female chastity and the monogamous marriage to preserve the family so necessary to its survival. The restrictive morality imposed by the bourgeoisie also helped maintain social distance. Once in power, Reich argued, the bourgeoisie "had to barricade itself against 'the people' by moral laws of its own, and thus imposed increasingly greater limitations on the primitive sexual needs of man." [11]

While the bourgeoisie was working out its own rigorous sexual morality, the working classes were beginning to revise their life styles along similar lines. The lower classes have not left a multitude of literary sources, but there is sufficient material to enable historians to conclude that the values and ideas of the upper classes filtered down the social scale, although they were undoubtedly interpreted and acted upon in a radically different spirit at the lower social levels. Maurice Quinlan has given a favorable evaluation of the effect of the moral reform of the early nineteenth century upon the lower classes. He concluded that "the low-lived, brutish masses of the eighteenth century were being transformed into the more decent, law-abiding common people of the Victorian era." [12] Steven Marcus has argued that the typical Victorian values such as "chastity, propriety, modesty, even rigid prudery" had a positive effect on the urban lower classes of the Victorian era.[13]

The Dutch historian Jos van Ussel has argued that the bourgeoisification of society is responsible for the "typical Western antisexual ideology. The more the bourgeoisie progressed, the more was sexuality repressed." This morality involved the transformation of the human body from an instrument of pleasure into an instrument of production, which hindered the full enjoyment of sex. It emphasized self-reliance, self-control, and the love of work—all of which inhibited a sexuality involving mutuality, abandon, and playfulness. The bourgeoisie was also responsible for an "interiorization" of sexuality. In contrast to the more public indulgence in sexuality of the Middle Ages, in the bourgeois home sex was carried out behind closed doors, under covers, and in darkness. The bedroom was set off from the rest of the house. This led Van Ussel to remark that "the intensification of the sense of shame has followed from the privatization of the bedroom." [14] He concluded his comment on bourgeois sexual morality by registering dismay that the Brownings never saw each other naked.

The stirrings of feminism in the first decades of the nineteenth century stimulated a defensive reaction among men who sought to reaffirm the subordination of women within the family. Although the argumentation over male-female relations fired up again in the latter part of the century with more virulence and more effectiveness, the pamphlets of such early feminists as Mary Wollstonecraft served to remind men that their privileges were at least open to question.[15] The "ideal of chastity" and the rule of feminine subordination were added to the edifice of Victorian sexual morality, and the feminist movement had to recoup for half a century before resuming its campaign for sexual, as well as political and economic, equality. In a recent study, John and Robin Haller concluded that the women of the Victorian era accepted the restrictions imposed by the severe sexual morality to achieve a measure of independence within the family. Though the cost was high in terms of sexual frustration, the women of this period observed strict continence even in marriage. The Hallers challenged the idea that Victorian sexual morality had a negative effect upon everybody who was forced to observe it. "Until contraception could be taken from the streets and used in the home, and until the woman won the right to greater mobility outside the home circle, her effort to define—and restrict—her sexual role in marriage was the only sexual freedom available to her, and her behavior should be considered an important step in the transition to modern womanhood." [16]

Peter T. Cominos makes a connection between bourgeois sexual morality and capitalistic thinking. An essential precondition for capitalist expansion was the accumulation of capital, which required the early entrepreneurs not to spend their money on pleasure, but to reinvest. The drive to save and reinvest, Cominos has argued, carried over into the sexual sphere. Throughout the nineteenth century in England a sexual orgasm was described as "spending." Men were therefore admonished not to "spend" too much of their sexual energy in order to be able to enjoy what they believed to be a normally regulated sex life *and* to have enough energy left over to perform the tasks of everyday life. "Continence in sex and industry in work were correlative and complementary virtues. The respectable Economic man must not be the sensual man who had failed to conquer himself, but the Respectable sublimated sensual man." [17] Lewis Mumford anticipated Cominos's interpretation of bourgeois sexual morality when he wrote that sex lost its intensity among the middle classes in the nineteenth century. "The secrets of sexual stimulation and sexual pleasure were confined to the specialists in the brothels. . . . Moralists looked upon it as a lewd distraction that would take the mind off work and undermine the systematic inhibitions of machine industry. Sex had no industrial value." The starvation of the senses among the prosperous middle classes, he concluded, "created a race of invalids: people who knew only partial health, partial physical strength, partial sexual potency." [18]

Christianity's contribution to the restrictive sexual morality of this period has been explained as a consequence of the asceticism that accompanied the rise of evangelical Protestantism in Europe and America during the early decades of the nineteenth century. This movement gave the moralists some of their most effective arguments for the need to severely limit sexual pleasure. E. P. Thompson has discussed the impact of Methodism in England on sexual morality. "Methodism is permeated with teaching as to the sinfulness of sexuality, and as to the extreme sinfulness of the sexual organs." The male sexual organs in particular "were the visible fleshly citadels of Satan, the source of perpetual temptation and of countless highly unmethodical and . . . unproductive impulses." The Methodists, he concluded, turned the spontaneous human feelings into sources of evil: "Joy was associated with sin and guilt, and pain . . . with goodness and love. . . . To labor and to sorrow was to find pleasure, and masochism was 'love.' " [19]

In a 1908 article, " 'Civilized' Sexual Morality and Modern Nervous Illness," and again in a full-length study of 1930, *Civilization and Its Discontents*, Freud offered a psychoanalytic explanation for modern sexual morality. He maintained that the progress of civilization necessarily demands ever more instinctual renunciation. The first civilized act of a child—toilet training—involves a renunciation of the instinctual desire to relieve pressure in the bowels immediately and the imposition of a requirement to wait for the "proper" time and place. The institution of marriage further demands severe regulation of the discharge of sexual desires according to the demands of a monogamous society. But Freud's argument does not explain the historical picture, because the evidence of the last hundred years shows relaxation of the instinctual renunciation imposed on sex life. The argument he suggested must, however, be considered along with the others to explain the restrictions that were imposed on the sexual life of

Europeans during the last century, even though the gradual relaxation of them, which I shall trace, tends to contradict Freud's pessimistic comment on the future of human sexuality.

By 1850 Victorian sexual morality was firmly rooted in Western life and thought. It maintained a hold on sex life until well into the twentieth century, when developments in medicine and sexual hygiene, as well as social and cultural changes, demanded revision and relaxation. But old values die hard, particularly those concerning the treatment of the human body, and in 1925 Mencken could still get a hearty laugh from his definition of puritanism—"the haunting fear that someone, somewhere, may be happy." In the following chapters I shall survey a number of cultural developments which illustrate the gradual dissolution of that morality.

Chapter 2

The Sartorial Assault on the Body

A good social history of clothing has yet to be written, and it is not my intention in this chapter even to outline one.[1] I want merely to survey some of the major developments in fashion that reflect changing attitudes toward the human body. I assume that clothing is an extension of the body and an expression of body image, and that therefore a history of clothing offers a good index to attitudes about the body.

The clothing of the nineteenth century illustrates the extraordinary difficulty that Europeans of that period had in dealing with their bodies. In no other age throughout history was the human body, in particular the female body, so concealed and so disfigured by clothing. By contrast, men in their static and drab attire appear to have been doing penance for the extravagance of female dress.

The century began with good intentions with the light clothing inherited from the Directory. The corset was eliminated, and women were draped in airy whites in an effort to imitate the dress of antiquity. This style was intended to correct the artificiality of the dress of the Old Regime, but by the 1820s the assault on the female body began once again with the return of the corset. Hemlines began to drop, and women were buried under ever more complicated, expensive, and restricting clothing.

In the 1830s women's clothing began to cover up the body with increasing thoroughness. Skirts lengthened to the floor to conceal even the ankle, the body was buried beneath layers of petticoats, the bonnet was worn close to the head, shawls were used to cover the upper part of the body, and elaborate hairdressings were abandoned except for demure ringlets which framed the face.[2] In the 1840s fashionable colors became dark and somber as the female body was encased in the billowy and cumbersome crinoline.[3] The wire-framed crinoline appeared in 1856 and served to lighten women's clothing for the twelve years it remained in fashion. Though the crinoline concealed the woman's entire lower body, it also served to provoke interest in what was underneath—and at a time when women did not wear underpants. A number of observers have pointed out that the occasional swings the crinoline made in a high wind or upon entering a coach aroused erotic interest among male observers. The extreme danger of fire led to its final demise. Peter Quennell reports that one Duchess de Maillé was burned to death at a friend's fireside when her crinoline caught fire; the Archduchess Mathilde, when discovered smoking, attempted to hide her cigarette in her petticoat and went up in flames, and a French actress was burned to death on stage when she got too near a gas lamp. Another unfortunate woman wearing a crinoline was blown off the deck of a ship into the sea and drowned.[4]

The crinoline gave way to the bustle in 1868, and there was a shift of attention to the back of women's dresses. By the early 1870s the bustle had dropped down to the hips, and in 1874 it began to disappear altogether until it returned in 1881. Erotic interest in the backside was revived in the mid-1880s with a new bustle that protruded straight out from the small of the back. In the 1890s the bicycle forced women to abandon bustles altogether, but a thorough reform of women's clothing had to wait until the First World War, when both the under and the outer wear of women first responded to the demands for clothing reform that had gone unheeded throughout the previous century.

A striking shift in male attire occurred at the end of the eighteenth century, when men gave up the bright colors and form-fitting garments of the Old Regime aristocracy and adopted the dark and modest uniform of the bourgeois gentleman.[5] This austere attire remained largely unchanged throughout the century except for the brief vogue of the dandy, embodied in the attire of George (Beau) Brummel in the first decade of the century. His costume consisted of a blue coat which buttoned tightly over the waist with tails cut off just above the knees and with a high lapel which rose almost to the ears. He wore skin-tight pantaloons tucked into black boots. The most famous part of his outfit was his cravat, which he is reputed to have spent two hours arranging each morning.

A variety of interpretations have been offered to explain the historical significance of this one bright moment in nineteenth-century male attire. Carlyle attacked dandyism as a tendency of the aristocracy to waste time and energy on frivolities.[6] Baudelaire viewed the dandy as a rebel against bourgeois materialism. "Dandyism appears especially in transitory epochs when democracy is not yet all powerful, when the aristocracy is only partially tottering and debased. . . . Dandyism is the last burst of heroism in decadence."[7] And at the end of the century Max Beerbohm commented with approval that the dandy *is* all exteriority; he expresses his essence, Beerbohm argued, solely through fashion.[8]

Aside from this one excursion into sartorial daring, male costume remained simple, reasonably functional, and stable. Critics of fashion repeatedly complained about the lack of color in male dress.[9] If I am correct in viewing female attire as an attack on the body as much as an effort to conceal it, male attire appears to have been motivated by the desire to ignore the male body through the use of subdued colors and conservative tailoring. Toward the end of the century George Bernard Shaw commented on the pathetic failure of man's dress. "The great tragedy of the average man's life is that Nature refuses to conform to the cylindrical ideal, and when the marks of his knees and elbows begin to appear he is filled with shame."[10] The male "cylindrical ideal," his white linen, his unerotic tailoring, and his insistence on colorlessness were manifestations of a flight from corporeality. The bourgeois gentleman of the nineteenth century did not want to enter the race for sexual selection by even acknowledging, let alone embellishing, his corporeal existence. He preferred to secure a wife by offering stability and a model of emotionally austere behavior. However, he did not allow women to remain in as subdued attire as he chose for himself, and women were caught between the contradictory demands to be both physically desirable and morally proper. One article of female underclothing perfectly illustrated these contradictory demands—the corset. It restricted her natural movements and general physical well-being, but at the same time was conceived to accentuate her sexual appeal even at the risk of injuring her sex organs.

The first serious effort to reform women's clothing came in 1849 when an American, Mrs. Amelia Bloomer, introduced "bloomers" or trousers for women to wear under their dresses. The purpose of this garment was to free women from the welter of undergarments required to fill out large skirts and offer some kind of cover for their legs. But the bloomer failed and was not revived until the end of the century, when bicycling led to its popular acceptance.

During the latter half of the nineteenth century the call for reform of women's clothing became steadily more vociferous as physicians, aestheticians, and physical culture advocates studied the destructive consequences of tight lacing. In 1856 the Englishwoman Roxey Caplin opened an attack on at least certain kinds of corsets. She rejected the idea of eliminating the corset entirely but insisted that it must be improved from a "scientific" study of the body. She then listed twenty-three different kinds of corsets suited to different needs: for infants and young girls, for the corpulent, for girls who grow too rapidly, for pregnancy, even to prevent children from standing on one leg. She concluded by reminding women that stays were necessary to prevent sagging of the breasts, and she offered the example—which no doubt struck terror in the hearts of many women—of African women whose breasts hung so low that they could nurse infants behind their back by flinging their breasts over their shoulders or under their arms.[11] In 1885 another English reformist recommended lighter clothing for women which would permit freer perspiration through the skin and easier movements of the body. "It is no rare thing," she wrote, "to meet ladies so tightly laced that they cannot lean back in a chair or sofa; if they did so, they would suffocate."[12] Although she was strongly critical of the corset, she recognized that women could not give up tight lacing overnight, and so she offered instructions for slowly reducing the tension.

The German fashion critic Emanuel Hermann surveyed the

"natural history of clothing" in 1878. He mentioned a number of the problems created by contemporary dress, which included a special concept, *"ängstliche Reinlichkeit"* (anxious cleanliness), an emotional state which he believed was caused by the masses of underclothing worn by women. The Greeks valued the human body as a gift of the gods, Hermann said, but current morality viewed it with scorn. "There's not a single beautiful line of the body that is not cruelly crossed and cut up by our clothes." [13] We must begin by investigating the beauty of our bodies and insist that clothing be made to enhance it. Tailors should make clothes according to the laws of human form and movement, which express the beauty of the body. These principles were restated by numerous reformists in the following decades.

An important contribution to the campaign for clothing reform was made in 1901 by Paul Schultze-Naumberg. The title of his study, *The Culture of the Female Body as a Foundation for Women's Clothing*, reveals the sequence of his analysis from anatomy to aesthetics. For centuries we have had an incorrect conception of the body which has recently degenerated into the artificial and decadent aesthetic sensibilities that overvalue large breasts and small waists. To conform to these ridiculous standards women corset themselves before eating, breathe incorrectly, and, in some extreme instances, have their lowest rib removed. After years of wearing the corset, the decorseted woman is astonishingly ugly. Her posture is twisted, her skin is discolored and sometimes marked by bloody streaks. Tight lacing leaves a permanent mark just above the navel. Schultze-Naumberg profusely illustrated his work with photographs showing permanent disfiguration caused by the standard tight lacing (fig. 1) and by the special "health" corsets introduced around 1900 that gave the female torso a characteristic "S" shape (fig. 2). He argued that corsets impaired the health of almost every vital organ, and he included several line drawings illustrating the deformations of muscles, bones, and internal organs caused by them. Constant tight lacing prevents the diaphragm from functioning and restricts respiration to the short and shallow breathing in women that produces vertigo and often causes "vapours" or fainting spells. The kidney and bladder are squeezed, causing a variety of urinary disorders. Digestion is disturbed, and constipation is a constant problem. The sex organs are often so badly atrophied that childbirth is unnecessarily painful and sometimes dangerous. The entire musculature of the abdomen never develops properly, with the result that women cannot exercise properly and often have trouble regaining their shape and muscle tone following childbirth. In the face of these numerous dangers, clothing reform, and in particular the elimination of the corset, is of vital importance. We must return to study the naked body. We need a "completely new understanding of the beauty of the human body . . . a new concept of corporeality." And from that concept a more functional set of clothing must be derived.[14]

Schultze-Naumberg's contribution to the literature on clothing reform is an excellent summary of the major arguments made in the closing decades of the century, but is of interest for another reason: his future career in the Nazi regime. I shall recount his activities on behalf of Nazi culture in Chapter 17, but here I should mention that his early concern for clothing reform reveals signs of the racist and nationalist ideology that would become more prominent in his later years. Schultze-Naumberg's career is not unique in German cultural history. A number of radical reformers who enlisted in the causes of clothing reform, the physical culture movement, and the youth movement were motivated by a hatred of what they believed to be the decadent and artificial accretions of the urban and industrial society that was coming to dominate German life. They believed that the return to nature and health was a necessary part of a movement to return to a uniquely German and Aryan past.

The career of Christian H. Stratz parallels that of Schultze-Naumberg and reveals a similar pairing of reformist and conservative ideas. Stratz's early publications were on gynecological anatomy. In the early 1890s he became the first European gynecologist to visit and work in Java. He recorded being struck at the superior physical condition of Javanese women, particularly their muscle and skin tone and the excellent condition of their reproductive organs, which he attributed to the fact that they did not wear corsets and were able to exercise and expose their bodies to fresh air and sunlight. In 1897 he published a comparison of the anatomical and aesthetic differences between European and Javanese women.[15] In 1900 Stratz began to support the movement for reform of women's clothing. His work as a gynecologist enabled him to understand precisely how destructive the corset was for the female reproductive organs. His book *Women's Clothing and Its Natural Development* was a mixture of hygienic, moralistic, and artistic theorizing, all subordinated to a strong racist and nationalist appeal for the preservation of racial superiority. It is an excellent example of the fusion of progressive-minded appeals for the physical culture movement and reactionary politics so frequently encountered in this period. He argued that the naked European woman was a nervous, unhealthy, asexual hybrid with a boyish body and a depraved soul; that a normally built woman with breasts and hips could not fit into the clothing required by high fashion. He concluded his evaluation with a passionate appeal for eliminating the corset, which he claimed was threatening the racial superiority of the European woman.[16]

While women's clothing was a health hazard, there was a crying need to reform male attire for aesthetic reasons. One German critic compared women and men to "bright flowers surrounded by swarming beetles." The stiff collar, the predominance of black, the cylindrical top hat and appendages all came under criticism. The German fashion critic Fritz Müller complained that men wore "oven pipes on the legs," "jackets that yawn from boredom," "collars that crackle from starch," and "hats that rise out of the head like stiff bubbles." [17] But in spite of all the calls for reform, male clothing would also have to wait until the war to achieve any significant changes, and even then it did not change as much as did women's apparel.

The First World War had a revolutionary impact on European morality. I shall explore many of its sexual consequences in a later chapter. The war also inaugurated changes in clothing which finally began to respond to the advocates of reform, who had been unheeded for the past half century. The exigencies of a wartime situation led to a reduction of interest in high fashion. Shortages of materials forced designers to use less, and dresses began to shorten. In England dresses were six inches off the floor by February 1915.[18] A shortage of manpower obliged women to take on jobs that had traditionally been reserved for men, with the result that women employed in factories in England were forced to abandon the corset, the petticoats, and the puffy leg-o-mutton sleeves and adopt safer clothing to work around the machines. The change in jobs followed a change in sex roles, which led to a masculinization of dress for women, so that by the end of the war women were seen wearing men's coats and ties and a variety of modified military and civilian uniforms.

The war finally brought the end of tight lacing. The shortage of metal discouraged manufacturers from producing corsets with metal stays, the demands of women's new employment made wearing corsets impossible, and the emergencies of a wartime situation provided a favorable psychological climate for adopting rational dress. The brassiere was introduced in 1912 and began to become popular around 1916.[19] Subsequently the corset was shortened to the dimensions of the present-day girdle, allowing freer movement of the entire body. One historian concluded that the war "allowed women to get rid of the bastions of nineteenth century respectability around their legs and thighs." [20]

In the decade following the war, women took revenge on each of the salient parts of the costume that had kept them imprisoned for the past century. For the fleshy feminine ideal of the nineteenth century they substituted a lean look. Upright posture and athletic bodies replaced the fashionable helpless "spineless look" that one fashion magazine, *Vanity Fair*, had celebrated in 1913. The hourglass or "wasp-waist" figure gave way to a boyish look, and the waistline disappeared altogether. The breast was de-emphasized by flatteners, and hemlines crept higher until they reached just below the knees by 1928. Englishwomen cut their hair ever shorter, first with the "bob" and "shingle" and then with the most severe "Eton crop." The new woman, the "flapper," challenged the sexual subordination to which women had been subjected in behavior as well as appearance.

The military uniform was another casualty of the war. For a century it had been an integral part of the masculine ideal, but four years of slaughter had permanently changed the status of the uniform and with it the link between manliness and military bearing. Hermann Broch has explored the psychological function of the uniform in *The Sleepwalkers*, a novel about a Junker officer Joachim von Pasenow.

> A uniform provides its wearer with a definitive
> line of demarcation between his person and the world.

> . . . it is the uniform's true function to manifest and ordain order in the world, to arrest the confusion and flux of life, just as it conceals whatever in the human body is soft and flowing, covering up the soldier's underclothes and skin. . . . Closed up in his hard casing, braced in with straps and belts, he begins to forget his own undergarments, and the uncertainty of life.[21]

At times von Pasenow wished that his uniform were part of his skin; he took comfort in the concealing, supporting, defining function of his uniform and grew to feel ashamed and exposed when in civilian clothing.

In the postwar years, as men began to work out a new masculine ideal, even their gait and posture changed. One historian has observed that "the straight-as-a-ramrod pose became old-fashioned, and a relaxed, slightly drooping posture was considered the thing," although some colorful die-hards were still to be seen in Vienna in the 1930s leaning slightly to the right as they had done before the war when a sword swung from the left.[22]

Contemporary literature began to interpret the psychological and social significance of clothing. The initial inspiration for exploring its philosophical significance came from Thomas Carlyle in his strange essay of 1831, *Sartor Resartus* [literally, Tailor Retailored]. Advising his readers of the pioneer nature of his work, he began: "It might strike the reflective mind with some surprise that hitherto little or nothing of a fundamental character, whether in the way of philosophy or history, has been written on the subject of clothes." [23] The German mind, he insisted, was especially suited to such an investigation, and he then proceeded to quote from a fictitious book by a Professor Teufelsdröckh (Devil's Filth) which commented on the moral, political, and religious significance of clothing throughout the ages. The essay was essentially a critique of German academic scholarship, but occasionally it touched on its putative topic—the philosophy of clothing. The first purpose of clothing, Teufelsdröckh argued, was not protection or modesty, but ornamentation. There followed a long discussion of the social function of clothing to establish social rank. But after announcing with such fanfare his intention to examine the philosophy of clothes,

Carlyle digressed repeatedly to comment on a variety of foibles of his time.

The German psychologist Hermann Lotze offered an early interpretation of clothing as an expression of body image. "Clothing, by adding to the apparent size of the body in one way or another, gives us an increased sense of power, a sense of our bodily self—ultimately enabling us to fill more space." It is easy to understand how he could have made that particular formulation in an age of epaulettes, padded shoulders, padded calves, long trains, and the crinoline.[24]

In 1899 Thorstein Veblen offered a biting critique of the social function of dress as a means of conspicuous consumption in a pecuniary culture. Bourgeois clothing at the turn of the century, he argued, was intended to demonstrate that "the wearer can afford to consume freely and uneconomically . . . [and] that he or she is not under the necessity of earning a livelihood." The charm of patent-leather shoes, stainless linen, and the walking stick is that they advertise the wearer's idle and rich social status. A woman's clothing demonstrates her uselessness even more blatantly. The bonnet, high-heels, long hair, and wide skirts show that she does no labor, while the corset is "a mutilation, undergone for the purpose of lowering the subject's vitality and rendering her permanently and obviously unfit for work."[25] The German sexologist Eduard Fuchs added to Veblen's critique of the motives underlying some of the more ostentatious attire of the bourgeoisie. The plunging neckline has come to be worn in bourgeois society only on special occasions when the other classes are excluded, such as at evening parties. This contrasts with the aristocratic tradition of allowing women to appear everywhere extremely décolletées. Fuchs interpreted this change as a further effort of the bourgeoisie to set itself off from the lower classes by prohibiting them from seeing the exposed part of their women's bosoms and of reserving that privilege only for members of their own class.[26]

In a psychoanalytical study of clothing in 1930, J. C. Flugel concluded that the movement away from the artificial was a sign of progress. "This increasing satisfaction in the more natural forms of decoration and the corresponding distaste for the grossly artificial seem to imply that human beings, as they advance in culture, become, on the whole, more prepared to accept the human body as it is, more inclined to find beauty in its natural shape."[27] The positive evidence I have collected on changing attitudes toward the body supports this. But it is sobering and disconcerting to note that it took the insane destruction of human life in the First World War to transform the demands of reformers into concrete historical change and bring about the beginning of the world of fashion's acceptance of the human body.

Chapter 3

The Body in Art

The history of the nude parallels some developments in the history of clothing and reveals a similar growing willingness to explore and accept the corporeal side of human existence.[1] The depiction of the unidealized body was not an invention of the nineteenth-century Realists. In the seventeenth century Rembrandt painted the wrinkles and sagging of aging bodies, while Rubens portrayed corpulent women bending and twisting in a vast panorama of human flesh. Boucher and Fragonard in the eighteenth century painted lusty wenches and provocatively posed courtesans, but in the late eighteenth century the artistic guidelines of neo-classicism prescribed the idealized bodies such as those of David and Gérard. Their subjects were generally young, healthy, and attractive, and were carefully posed to highlight only the most flattering surfaces and forms (fig. 3). Though this rendering of unblemished bodies was in some respects a welcome alternative to the view of the body as a vessel of sin as portrayed by some earlier religious painters, and perhaps was intended to illustrate the eighteenth-century ideal of the perfectibility of man, it also represented a flight from some obvious facts of human corporeality.

In the course of the nineteenth century, artists gradually challenged the classical ideal and ventured to portray nudes in unidealized physical condition, settings, and poses. To some extent they were merely recapturing the artistic freedom of the seventeenth and early eighteenth centuries, but in so doing they explored the body as subject matter with particular daring. In this chapter, I will follow that development which culminated in the nude studies of the Post-Impressionists and Expressionists.

Ingres's nudes reveal a number of conflicting attitudes toward the body among artistic circles of the early nineteenth century. His nudes show a growing tendency to deviate from classical proportions, although they never cease to be highly idealized. Though some of his later canvases are explicitly erotic, he scrupulously observed the convention of not rendering pubic hair. Instead of using the classical device of posing the figure so that she would not show any pubic hair, he painted several with the pubic region in full view (figs. 4, 5), but covered by unbroken pink flesh. His *Grande Odalisque* (1814) is a mysterious mixture of emotional restraint and sensuality (fig. 6). The flesh is molded and perfected with care, but appears to be untouchable. Any intrusion would disturb the careful composition of the subject surrounded by artfully draped fabrics of gold and blue. This famous nude contrasts sharply with the blatant sensuality of Goya's *Naked Maja* (1802), which appears in the history of art as a forewarning of the explicit eroticism that the artists of the remainder of the century would explore more deeply.

In 1862 the aging Ingres capped a lifetime preoccupation with feminine sensuality with *The Turkish Bath* (fig. 5). The critic John L. Connolly, Jr., has called it "the masterpiece of a lifetime search to create an allegory of the senses wherein the spectator is the living personification of sight." Connolly surveyed a number of Ingres's nudes to demonstrate how Ingres systematically tried to represent the senses in his art. In *Turkish Bath* all five senses are represented once in the foreground and once in the background. The food in the foreground and the eating in the background represent taste; sound is taken up with a stringed instrument and a tambourine. A censer applied to one of the major female subjects in the right foreground compliments a censer in the far left background to include the sense of smell. Touch is portrayed by the fondling of the breast at the right and the woman being tickled with a feather in the center of the background. The viewer supplies the visual sense for the entire scene of undulating female bodies.[2] This effort by Ingres to treat each of the five senses points ahead to the systematic exploration of the senses in the decadent novel of Huysmans, *Against Nature*.

Several nude studies of Delacroix and Géricault around 1820 anticipated the realistic nudes that appeared during the latter half of the century. Though a number of Delacroix's studies of nude men and women explicitly revealed pubic hair, his finished paintings observed the convention of keeping the pubic region either concealed or shadowy. The bodies in Géricault's *The Raft of the 'Medusa'* (fig. 8) are depicted at a moment of extreme excitement—the men on the raft who had been drifting at sea are waving frantically at a ship that has appeared in the distance. The painting is a monument to the ability of man to survive severe physical hardship, but Géricault did not shrink from including the few who lost in the struggle and died at sea. Their lifeless bodies lay sprawled out conspicuously in the awkwardness and contortions of death. Géricault made studies of dismembered limbs and visited a morgue where he sketched corpses before he worked on the painting. Observing the conventions about pubic hair, he painted on the figure in the foreground a cloth draped over the genitals, which otherwise would have been visible. His *Study of a Nude Male* (fig. 9) shows a mixture of romantic and realistic conventions. The man's body is well developed, and his imposing stance shows a strong influence of the classical tradition. It is still far from the bending laborers of Millet or Van Gogh, but the body is evidently that of a worker with calloused hands and reddened elbows and feet. As it was a study, Géricault was able to draw explicitly the genitals and pubic hair.

In 1849 Courbet boldly challenged a number of "Romantic" sensibilities with *Burial at Ornans*. His challenge to the idealized nude came several years later with two highly controversial major paintings—*The Bathers* (1853) and *The Painter's Studio* (1855). These paintings announced what became the Realist movement in art, and the human body began thereafter to be painted in more natural settings and poses. But Courbet's work was transitional, because however much he intended to paint realistically he could not entirely abandon certain conventions of the classical ideal.[3] His *Bathers*, for example, still presented the major figure in a studied pose and draped with a towel, although he did mock the classical convention of draping the nude figure by having his bather dry herself with a bath towel. Also, the Greeks did not portray their female models from the backside, twisting awkwardly to show masses of posterior flesh. Courbet, as he tried to carry out his own injunction to paint objects as they were—be they landscapes, workers, or nudes—was making a major shift to a subject matter that painters would treat with increasing daring as the century drew to a close.

Two nude paintings of the same title by Courbet and Ingres illustrate the changes that were taking place around mid-century. Ingres's *La Source* (fig. 4), painted in 1856, shows a strict conformity to classicism and offers a highly idealized female body polished to perfection. Courbet's nude of 1869 shows the puffy flesh of a woman casually turned from the viewer (fig. 11). In accord with the Realists' stated intention to portray the world as faithfully as possible, Courbet rendered the woman's puckered left thigh—a flaw that Ingres no doubt would have eliminated.

Manet upset the French academy with two paintings in 1863, *Picnic on the Grass* and *Olympia*, both of which portrayed nudes who eyed the viewer frankly (figs. 10, 7). The unclad woman in *Picnic* provoked outcries from the artistic community because, it was argued, the presence of the fully clothed men made her naked rather than nude. Though there is a general casualness about the scene, some objected that the situation suggested that the two couples might intend to indulge further in physical gratification after lunch. Although Manet derived the composition and the theme of clad males with unclad females from a Renaissance painting by Giorgione, the subject matter seemed to offer a scene of earthy carnality. It must be remembered that Manet attempted to exhibit this work when Victorian sexual morality was at its peak. Under such conditions spectators could possibly have been sexually aroused by it—and as aestheticians had repeatedly insisted, sensual stimulation was not the function of true art.[4] *Olympia* was blatantly provocative. This nude courtesan stretches out on a bed and stares challengingly at the viewer. The black cat gives the scene a strong sexual connotation, the presence of the maid intensifies the voyeuristic suggestiveness, and the flowers the maid presents to her obviously have come from one of her many admirers. There is one touch of modesty in the painting: her hands conceal her pubic hair.

The problem of how to deal with pubic hair in nudes is one that the Realists could not manage in keeping with their demand to paint the world as they found it. One art historian has written, "Neither Courbet nor Manet ever faced squarely the Realist's dilemma, in representing the nude, with respect to pubic hair versus the classical (or hairless) alternative, and *that* versus the true representation of the female anatomy." [5] Throughout the nineteenth century artists depicted the pubic region in sketches and even small studies in oil (fig. 9), but never in major Salon paintings. Then toward the end of the century artists began to depict it more regularly. Van Gogh represented pubic hair in a nude of 1887, as did Toulouse-Lautrec and Aubrey Beardsley in the 1890s. In the twentieth century the Expressionists accentuated the human body and its sexuality by depicting the hair, the folds, and the blemishes that appear on it. As late as 1912 an exhibition of Egon Schiele's paintings was closed by the police because some of the paintings showed pubic hair.

Before considering the Post-Impressionists and Expressionists, I must first discuss a development in aesthetic theory that introduced the ugly as suitable subject matter and profoundly influenced the art of the late nineteenth century.

In a literary manifesto of 1827 Victor Hugo argued that art may include the ugly as well as the beautiful. The epic poets of antiquity limited their subject to "a certain type of beauty." The complete modern poet must realize "that everything in creation is not humanly *beautiful*, that the ugly exists beside the beautiful, the unshapely beside the graceful, the grotesque on the reverse of the sublime, evil with good, darkness with light." [6] The beautiful is a limited, simplified view of the natural world, while the ugly includes the great variety of forms found in nature and supplies an endlessly rich source of poetic images.

The German philosopher Karl Rosenkranz elaborated upon this aesthetic in a full length study of 1853, *Aesthetics of the Ugly*.[7] He justified his theory by arguing, Hegelian style, that we come to know things by studying their negation as well as their positive form. Just as the physician studies disease and the moralist studies evil, so the aesthetician must study ugliness, which, he insisted, is inseparably linked with beauty. Following sections on "natural ugliness" and "ugliness in art" he concluded with a discussion of the deformation of natural physical beauty in caricature, which he believed to be the most fruitful use of the ugly in art.

In 1857 Baudelaire published a collection of poems, *Flowers of Evil*. Its exploration of the world of evil and the grotesque shocked the French literary world. The title itself was disturbing, and one poem took up in detail the argument Hugo had made thirty years earlier. "Hymn to Beauty" was an invitation to follow artistic impulses whether they come from "Satan or God," whether they be "holy or vile."

> Do you come from the heavens or from the abyss,
> O Beauty? your gaze, infernal and divine
> Pours a mixture of blessing and crime.[8]

While this poetic manifesto announced the revolution in subject matter, another poem carried it out. "A Carrion" described a dead animal in terms usually used for natural beauty and human love. The poem is addressed to the poet's lover as the two walk by a carcass on a hot afternoon. Baudelaire contrasts her sweet living flesh with the decaying flesh of the carcass. The poem is a telling comment on love, aging, and death.

> Do you recall the object that we saw, my dear,
> That beautiful summer morning so gentle:
> At the bend in the path a vile carrion
> On a bed strewn with pebbles.
>
> Its legs in the air, like a lubricious woman,
> Burning and sweating poisons,
> Opened in a nonchalant and cynical manner
> Its belly full of exhalations.
>
> The sun shone on that rottenness,
> As if to cook it to a turn
> And to render a hundredfold to great Nature
> All that she had joined together.
>
> . . . .
>
> The flies buzzed about that putrid belly,
> From where battalions of black

> Maggots poured like a thick liquid
> Along its living rags.

The final lines address his charming companion, whom he envisions in a similar condition.

> You will someday be like that filth,
> Like that horrible infection,
> Star of my eyes, sun of my nature,
> You, my angel and my passion!
>
> Yes, so will you be, oh queen of charm,
> After the last rites,
> When you will be under the herbs and the flourishing grasses,
> To go moldy among the bones.
>
> Then, oh my beauty, tell the vermin
> Which will eat you with kisses,
> That I have preserved the form and the divine essence
> Of my decomposed loves! [9]

Following Baudelaire's bold lead, artists began to explore the morbid and the ugly.[10] The nude studies of the late nineteenth century were in part influenced by his poetry and by the mid-century revolution in the aesthetics of the ugly.

Degas, Toulouse-Lautrec, Rodin, and Rouault followed the lead of Courbet and Manet and completed the revolution in the depiction of the nude. Degas was intrigued by the active body and produced hundreds of studies of ballet dancers. He also painted more earthy women working, bathing, drying their feet, picking at their toes—all in striking contrast to the studied poses of the earlier period. Degas was fascinated with the body as it moved and did not limit himself to portraying only perfectly formed bodies. His female models are shown in a variety of broken-down states, in many instances brought on by overwork, drugs, or prostitution.

Toulouse-Lautrec chronicled the life of Parisian prostitutes, can-can dancers, and barmaids. His *Woman Pulling Up Her Stockings* (1894) is a frank study of an aging woman performing a mundane daily function (fig. 13). Her body is bent awkwardly, her pubic hair forms a visual focal point and the flesh folds, interrupting at every juncture the graceful lines that had for so long dominated the artistic conception of the female form. Some of Rodin's nudes show how completely artistic sensibilities had varied from the earlier insistence on "beautiful" subjects. *She Who Was Once the Helmet-Maker's Beautiful Wife* (1885) accented the angularity of the aging female form (fig. 12). The eye is led downward by the sagging of her flesh and the determined downward thrust of the pose, all by way of underlining the fact that physical condition declines with the passage of time. Rodin was also capable of creating human forms in the fullness of youth, but he did not avoid showing the way the body deteriorates in the later years. Between 1903 and 1907 Rouault produced what are perhaps the ugliest women ever to be painted: a series of studies of broken-down prostitutes (fig. 14). Kenneth Clark has commented on one of them: ". . . our dream of a perfectible humanity is broken by this cruel reminder of what, in fact, man has contrived to make out of the raw material supplied to him. . . . All ideals are corruptible, and by 1903 the Greek ideal of physical beauty had suffered a century of singular corruption." [11] Rouault's prostitutes have lost even the sensuality and eroticism associated with Toulouse-Lautrec's. His are merely sagging masses of flesh. His *Little Olympia* (1906) mocked the motif of the reclining nude in a final rejection of the classical ideal.

    The portrayal of progressive deterioration in the physical appearance of the nude in the works of the Post-Impressionists must be viewed as more than a declaration of war on traditional aesthetic conventions. It also foretells of a general movement, observable in the work of many artists and intellectuals of this time, to understand the nature of man and the world, especially when that understanding led to the rejection of the self-delusions that they believed had blocked understanding and artistic vision. And of course the artists exploded with indignation at the prior restraints as they came to understand that they could create great art which did not perfect the outer physical structure of man.

    Though this vision of the aging and decrepit female form did not entirely dominate painting, the Post-Impressionists and Expressionists revised the going conception of the nude so thoroughly that serious painters would now be most reluctant to use the classical feminine ideal. The last burst of idealization of the female before the turn of the century came in the prolific work of Renoir, with his rosy-cheeked, full-breasted women. His corpus is a celebration of the traditional woman—childlike and maternal, sensual and passive. They smile as if peering naughtily through the screen of propriety that had for ages shielded them from the gaze of men. Renoir invites his audience to a visual feast on the warmth and comfort of the female body.

Cézanne was among the first of modern painters to subordinate content to form. He gradually strained the content out of his paintings in a series of landscapes and still lifes that showed meticulous attention to the arrangement of colors and shapes. His nudes also revealed this new conception: they became more and more simplified as his work progressed, and eventually came to be the starting point of a purely artistic vision. His *New Olympia* (1872) satirized Manet's earlier canvas and rendered the body of the woman so that her details were entirely obscured. It gave merely an impression of a woman lying on a bed. His *Bathers* (fig. 16) shows further his diminishing interest in anatomy and the traditional artistic concerns about proportion and perspective. The nude bodies and the landscape in which they are assembled are subordinated to the overall pictorial vision. The shape of the bodies, the color of their skin, and the expressions on their faces are ignored.

After the turn of the century the Cubists followed Cézanne's lead and began to break up the nude body and rearrange its parts, again in deference to purely artistic requirements. Picasso's *Women of Avignon* (1907) is a pictorial Cubist manifesto announcing his intention to depart entirely from the tradition of painting the human body as it is naturally constructed (fig. 15). The artist is entitled to reconstruct the body as he wishes, and henceforth, at least for a certain group of artists following the Cubist school, the human body would be merely a point of departure, a source of artistic inspiration, to be subordinated to artistic purpose.

The Expressionists pioneered another important aspect of modern art—the portrayal of the artist's emotions. The traditional requirement to paint what was "out there" gave way with the Expressionists to the effort to paint the artist's inner response to the world. Among the German Expressionists in particular, the nude expressed the inner tensions and impulses that had been locked in the artistic mind. The Norwegian painter Eduard Munch pioneered this genre in the 1890s with studies of intense emotional states such as lust, jealousy, and anxiety.

The first group systematically to pursue this stylistic innovation was *Die Brücke*, which formed in Dresden in 1905. Its first four members—Ernst Kirchner, Fritz Bleyl, Erich Heckel, and Karl Schmidt-Rottluff—worked together in a rented butcher shop, which they converted into a studio where they did their nude studies. They sought to portray in particular the violence and intensity of sensuality.[12] Their lives and work were in sharp contradiction to what they believed to be the excessively restrictive sexual morality dominating the bourgeois class they so detested. "The *Brücke* . . . wanted to bring back the naked figure into landscape and create a harmonious composition in which the two merge and man becomes part of nature." [13] Another Expressionist group which formed in Munich in 1912, *Die Bläue Reiter*, was led by Wassily Kandinsky, whose theoretical statement *Concerning the Spiritual in Art* (1912) speculated about the relation between the senses. Colors suggest certain feelings, Kandinsky argued, and lines trigger specific emotions. Human sensuousness could therefore be put on canvas by finding the appropriate artistic representations. Following Kandinsky's theoretical lead, many Expressionists tried to portray their inner emotions by using the corresponding colors and lines.

The Viennese Expressionists explored the world of sexuality with a degree of daring that paralleled Freud's clinical findings made in Vienna at that time. Gustav Klimt's nude studies reveal a fascination with the human form, twisting and straining to attain new positions. The sexuality depicted in his work hints at the clutching and suffocating nature of sexual relations.[14] Egon Schiele's nudes portray the effect of hard work, malnutrition, and aging on the human body. His women confront the viewer with their legs casually spread with the abandon of tired work animals (fig. 17). One drawing depicts a pregnant woman with a swollen belly and darkened nipples. Another shows a nude man from behind with his scrotum displayed with bull-like prominence. One can almost smell Schiele's nudes. Their ribs and breast bones show through worn flesh, and their limbs are assembled in improbable and grotesque patterns. Schiele strains to emphasize the angularity of the human form and unmercifully confronts the viewer with its various functions. In two separate paintings of a *Blind Mother*, infants nurse at random points of their mother's body, suggesting that it is a mass of nursing flesh upon which a nipple can open at any point on the torso. The interlocking limbs of *The Family* (1918) suggest a biological intertwining of the appendages of its three members—father, mother, and child—with each one couched within the bodily frame of another. Schiele's conception of the human body anticipates Sartre's later emphasis on its stickiness, particularly in the relations between bodies. Schiele's nudes announce that the human body works desperately to maintain itself, being pathetically weak at the same time as it is impressively strong. The life of man, he implies, is more organic than spiritual, more corrupted by fear, routine, and disease than ennobled by love and tenderness.

In spite of his fascination with the animal nature of human sexuality, Schiele also strives to portray the nobility of even the most grasping and clinging sexuality. *The Embrace* (1917) marks a high point in the history of art for the representation of the totality of the sexual union (fig. 18). The couple appear to merge with the background as with each other. In the sexual embrace, borders can be obliterated and the world becomes for one brief moment unified and complete.

The invention of photography revolutionized conceptions of the human body and had a consequent impact on the nude in art. Manet's *Olympia* was influenced by the many pornographic photographs available at the time. Less than ten years after the production of the first daguerreotype, Delacroix acknowledged his debt to the photograph for his studies of the nude. One critic has concluded that "his previous perceptual *Gestalt* was formed by the Renaissance/Baroque tradition. It required the photograph to create a new visual conception of the human body." [15] Artists reacted to photography positively by trying to imitate the realism that could be captured with a camera, and negatively as they began to abandon the effort to achieve visual fidelity that had inspired their predecessors over the centuries. With the camera available to record precisely what could be seen, artists began to think that their function was to record their own subjective response to the world and then decompose it and reconstruct it in color and form. The wild distortions and exaggerations of the human nude that the Expressionists produced were made possible, in part, by the liberation from literalism that the introduction of photography made in modern art.

An early study of the impact of the motion picture on attitudes toward the human body was published by the Hungarian critic Béla Balázs in 1923. In an anticipation of the interpretations of Marshall McLuhan later in the century, Balázs speculated that the film was transforming the emotional life of twentieth-century man by orienting him away from words toward movements and gestures. "Mankind is once again becoming visible." A culture dominated by words is intangible, abstract, and overly intellectual, and it tends to degrade the human body into a mere biological organism. But this new language of gestures is giving rise to the longing for the "silent, forgotten, invisible corporeal man." Contemporary culture seems to be substituting the visible body for the abstract spirit that previously preoccupied artists.[16]

One example of the strong visual orientation brought about by the film industry is the reputation Marlene Dietrich made from her role in *The Blue Angel* (1930). From this one performance her legs became a cultural symbol. Only the mass distribution of the film could enable her to acquire such a reputation from a few scenes in which she crossed and uncrossed and then recrossed her legs while straddling a chair. Moving pictures alone could have made possible the immortalization of her legs so soon after an era that concealed legs under full-length dresses because they were too suggestive.

The history of art is profoundly influenced by many other considerations than the one I have focused upon, but one general development can be clearly discerned from the late eighteenth century on—the increasing willingness of artists to view and to recreate the human body in an ever greater variety of conditions and poses. This development was part of a general cultural development that began to gain momentum around 1850 and which progressed steadily for the next century. The nude studies of Degas, Toulouse-Lautrec, and Rouault as well as the sculpture of Rodin all reveal this adventurous and unrestrained exploration of human corporeality. This movement culminated in the work of the German Expressionists, who announced defiantly that any human body, no matter how distorted by age, abuse, or disease, could be a worthy subject for artistic treatment. As we shall see in the following chapters, this new artistic acceptance of human corporeality is reflected in a great number of formal and imaginative works of this period.

Chapter 4

Contagion and Human Relations

Developments in medicine and public health in the latter half of the nineteenth century profoundly affected the way Europeans came to view their bodies. Medical science began to explain what agents were responsible for the transmission of disease, and improvements in public health enabled increasing numbers of the population to keep clean. At the same time Europeans began to show an awareness of the workings of their bodies and the ways odors, exhalations, and microorganisms from others could affect them. This chapter will survey some of those medical advances and offer some speculations on the effect of those discoveries on the prevailing conception of human nature and human relations.

To a large extent the impressive developments in personal hygiene that were made by the mid-nineteenth century were dependent on political changes and technological improvements which increased the amount of good water available for each person and provided for efficient and hygienic removal of waste materials. One important result of these technological advances was an increase of attention to the body. The ability to keep clean generates a desire to remain clean which, if prolonged, becomes a psychological necessity. And the preoccupation with keeping clean did tend to focus attention on physical existence.

Historians of public hygiene tend to agree that around 1850

Europeans began to take an active interest in improving the sanitary condition of their lives. An American sanitary engineer, William Gerhard, commemorated a "Half-Century of Sanitation" in a speech delivered in 1899. "All large sanitary municipal improvements," he said, "date from the year 1850. Before this date the practice of bathing was not a general one, and was entirely confined to river and sea baths." Gerhard reminded his listeners that Europeans also began to improve their water and sewer systems around this time: Hamburg in 1842, Danzig in 1869, and Berlin beginning in 1870. The first modern sewer system was begun in Paris in the Rue de Rivoli in the 1850s. And the British began installing a new sewer system following the cholera epidemic of 1848. "England set in 1842 the example for municipalities in providing public baths for the people, and since 1850 the principal cities of the Continent, particularly in Germany, have imitated it." [1] René Dubos has more recently made a similar generalization: "Through the efforts of public-minded citizens . . . around 1850 society slowly began to take an active interest in a more salubrious life—clearing slums, eliminating filth, providing fresh air and abundant, clean water." [2]

The French appear to have been somewhat slower in adopting the habit of regular bathing. The French historian Guy Thuillier has concluded that by the turn of the century regular bathing was still rare in France. Thuillier quotes Vacher de Lapouge in 1897: "In the Catholic countries hygiene of the skin is almost unknown. In France most women die without ever having had a bath. The same is true for most men, excepting their military baths." [3] Thuillier adds that in the 1870s it was forbidden for students in certain convents to bathe themselves. Around 1900 it was common for mothers to let their children steep in their own feces and urine for up to ten days. By 1919 most of the people in the region he studied (the Nivernais) used soap only to wash themselves before going to church.

Provisions for adequate bathroom facilities in homes and in public places also began to be made around mid-century. The Crystal Palace Exhibition of 1851 pioneered the introduction of "public conveniences" in England. The *Official Report* of the exhibition acknowledged that their installation helped reduce "the suffering which must be endured by all, but more especially by females, on account of want of them." [4] Gerhard reported that prior to 1850 there were few houses fitted with facilities for providing ventilation of bathrooms. The widespread use of toilet paper began in the last third of the century.[5]

An important large-scale study of the problems of inadequate public health provisions was published by the Englishman Edwin Chadwick in 1842: *Report on the Sanitary Condition of the Labouring Population of Great Britain*. The introductory chapters described the horrors of open sewers running through the streets, dung heaps beside every door in the slums, contaminated water supplies, and the nauseating stench in houses and shops produced by inadequate ventilation. Chadwick concluded that the crowding and the filth were partly responsible for the "immorality" of the poor. The result was sexual promiscuity, incest, homosexuality, and prostitution. He warned "that these adverse circumstances tend to produce an adult population short-lived, improvident, reckless, and intemperate, and with habitual avidity for sensual gratification." [6] The general form of his argument—that crowding generates sensuality—offers another explanation for the growing awareness of the human body in the late nineteenth century. The lack of space in urban dwellings forced people into more intimate and more frequent physical contact, with the result of heightened body awareness. Chadwick adduced numerous examples of parents and grown siblings of both sexes all sleeping together in the same bed. The physical contact, the voyeurism, the body odors, and, later in the century, the fear of contagious germs, all contributed to this heightened sensuality that Chadwick found so perilous for the morality of the poorer classes in 1842.

Of all the general historical factors that make up the surge of interest in the body, it is the concentration of bodies into increasingly restricted space, owing to the spectacular growth of cities in the late nineteenth century, that is the most important. The history of privacy deserves a full-length study, and I must limit myself to speculating in broad terms about its impact on body awareness.

Although this concentration of people tended to force them onto one another with greater frequency, European society became more conscious of a desire for privacy precisely as living space became more and more restricted. An example of this desire for privacy, as I have mentioned, was the relegation of sexuality to the bedroom and its restriction by conventions requiring that it be kept under the covers and in the dark. One possible explanation for these two apparently contradictory developments is that perhaps the dominant middle classes at this time responded to the growing constriction imposed upon them by accepting privacy and secrecy about bodily functions and by insisting that concealment of sexuality was a positive virtue, a sign of respectability. The middle classes could then hail Chadwick's warning about the wicked sensuality generated by crowding and reassure themselves that their own expensive privacy was another sign of their superior moral character. Following Chadwick's reasoning, the movement to improve public hygiene by reducing crowding was linked with the desire to limit the sensuality assumed to dominate the life of the lower classes.

The progress and popularization of the germ theory of disease produced a profound alteration in attitudes toward the body and in human social interaction. The discovery of specific microorganisms which multiplied in the bodies of diseased people and then infected others intensified the fears of contagion that had arisen intermittently during epidemics. Throughout history isolation was a standard method of dealing with epidemics and was often enforced legally to insure protection of threatened communities, but in the late nineteenth century scientists began to speak of the danger from proximity to persons infected with non-epidemic diseases. "After 1877," writes René Dubos, "physicians, as well as the lay public, became obsessed with the thought of disease germs." [7]

This obsession was the consequence of a number of specific discoveries about germs that gained widespread attention following the discovery that puerperal fever was transmitted by the infected hands of physicians attending women in childbirth. Oliver Wendell Holmes speculated on the contagious nature of this disease in a paper of 1843, "On the Contagiousness of Puerperal Fever," and the theory was confirmed by the Viennese physician Ignaz Semmelweiss in 1847. However, the theory met solid resistance from physicians, because it became mixed up with the issue of midwifery. Women who bore children attended by midwives had a better chance of survival than those who were attended by physicians, because the physicians transmitted puerperal fever from patient to patient in hospitals.

The outbreak of a devastating cholera epidemic in 1848 triggered a series of studies which began to explore the cause of the disease. There was no doubt that cholera was transmitted from diseased people, but the mechanism of that transmission was not at first understood. By the late 1840s the dominant theory was that miasma (odors given off by decaying bodies) caused the disease. The two greatest English epidemiologists of the early nineteenth century, Southwood Smith and Edwin Chadwick, believed that the smell of animal refuse produced the disease. In 1849 John Snow showed that the disease was transmitted by water, but even he did not know precisely what in the water caused the disease. The miasma theory continued to hold sway until well into the 1880s, when a series of discoveries of the specific microorganisms causing diseases led to its rejection.

In 1866 Louis Pasteur demonstrated conclusively that a living organism was killing the silk worm, and in 1870 he began to study anthrax, which attacked sheep and man. In 1881 he prepared an inoculation which made sheep immune from infection. By the mid-1870s the work of Pasteur and Robert Koch had firmly established the germ theory of disease within the scientific community.[8] The most important discovery in this area was made by Koch when he isolated the tuberculosis bacillus in 1882 and, in the next few years, isolated the germs that caused cholera and diphtheria. The germ theory of disease was further substantiated by the adoption of the technique of antiseptic surgery developed by Joseph Lister in 1870.

By the late 1870s the European medical community was inundated with revelations about a host of germs that attacked and destroyed the human body. The broad cultural consequence of these discoveries was a fear of contagion, which in turn engendered a fear of contact with other people. William James made an extraordinary confession of this general fear when describing "the antisexual instinct, the instinct of personal isolation, the actual repulsiveness to us of the idea of intimate contact with most of the persons we meet." And in a footnote to this bit of speculation he wrote: "To most of us it is even unpleasant to sit down in a chair still warm from occupancy by another person's body. To many, hand-shaking is disagreeable." [9] One marriage manual published in 1897 counseled married people to avoid "the absorption of the exhalations of each other's bodies, the weaker being injured by the fact that the stronger is likely to absorb vital and nervous force." [10]

One disease in particular figured in a good deal of imaginative literature of the period. Tuberculosis extended the process of dying and tested the force of love against the fear of contagion. Writers explored these passions with stories of young protagonists faced with a premature battle against the fatal suffocations of the disease. The slowly dying tuberculars in Dumas's *Camille*, Puccini's *La Bohème*, Schnitzler's *Dying*, and Gorki's *The Lower Depths* struggled with these conflicts, and Thomas Mann wove them into one of the literary masterpieces of the twentieth century.

Mann's *The Magic Mountain* (1924) records a number of attitudes about disease and the body in the early twentieth century. The symbolic historical message of the novel is easily interpreted: Europe (symbolized by the community atop the Berghof) is diseased, constantly taking its temperature, awaiting the onset of the fatal collapse—the First World War. The condition of European society is revealed to the hero, Hans Castorp, who has come to a tuberculosis sanatorium high in the Swiss Alps to visit his ailing cousin Joachim. Hans plans to stay for three weeks but remains seven years. At first he is shocked by the preoccupation with disease, but gradually he succumbs to the strange reverence for it. To belong at the Berghof one must be diseased, and so Hans is tempted to "measure" for the first time. He discovers that he has some temperature, and the house physician persuades him that he is anemic and that he has a "rough place" on his lungs—"almost a bronchia."

The progress of Hans's physical disease is accompanied by discussions on the nature of disease. The man of the Enlightenment, Settembrini, rejects romantic fascination with disease: "We are to honor the . . . body when it is a question of emancipation, of beauty, of freedom of thought, of joy, of desire. We must despise it insofar as it represents the principle of disease and death, insofar as its specific essence is the essence of perversity, of decay, sensuality, and shame." [11] The nihilist Naphta counters by extolling disease as the inspiration for genius and the progress of mankind. Hans decides in favor of Naphta and the ethic of disease as he falls in love with the hopelessly tubercular Russian patient Frau Chauchat. "For love of her, in defiance of Settembrini, I declared myself for the principle of unreason, the spiritual principle of disease." [12] The preoccupation with disease leads each character to attend to minute changes in his bodily state: the sound of coughs, the consistency of expectorations, the flush of complexions, and the constant measuring of fever. The most popular topic for conversation is fever—its subtle variations according to changes in diet, the time of day, and weather conditions. Hans's fascination with the body leads him to plunge into a systematic study of human physiology, organ by organ. Even his love for Frau Chauchat is intensified by his fascination with the mysterious disease that is relentlessly destroying her. His first confession of love, which he communicates in French—a language reserved for love and disease—reveals the extent to which love, disease, and death had become associated in his mind. "The body, love, death are one and the same. The body is both illness and voluptuousness. The body causes death; indeed, love and death are both sensuous—that is their terror and their great magic! . . . Love of the human body is the high point of humanitarianism, a greater educational force than all the teaching in the world!" He then describes the pleasure of viewing and caressing various parts of the body before turning to address Frau Chauchat in a final burst of medical sensualism: "Yes, my God, let me breathe in the odor of your knee-joint, beneath which the ingenious jointed capsule secretes its slippery oil! Let me devotedly touch with my mouth the femoral artery which pulses at the base of your thigh and which separates further down into the two arteries of the tibia! Let me smell the odor of your pores and feel your downy coat—a living vision of water and albumen, destined for the anatomy of the tomb, and finally, let me die with my lips on yours." [13] The one souvenir he keeps of Frau Chauchat is an X-ray of her tubercular lungs.

Just as the intensity of love and life are accentuated by the presence of disease at the Berghof, the moment of death gives life its final meaning and lasting judgment. The heroes of this absurd struggle with disease are those whose deaths are accomplished with dignity, but Mann takes equal relish in describing at length the panic of "the dance of death" of those who flail about in agony before the final convulsive suffocation. *The Magic Mountain* provides a dramatic catalog of a number of the fears that plagued Europeans at a juncture in history when medical science provided ever new hope for diagnosis, treatment, and cure, but at the same time remained largely helpless before the oddly shaped infectious germs that were communicated from one person to another in mysterious ways.

In addition to the fear of the killer diseases tuberculosis, cholera, and diphtheria, the contagious transmission of venereal disease added to the worries of Victorian men and women. The history of syphilis is as confusing as the disease itself. To this day a debate rages over the cause of the apparently simultaneous introduction of syphilis into both Europe and the New World following Columbus's first voyages. Did Europeans contaminate the Amerinds, did both somehow introduce strains of syphilis to which each of the two respective societies were not immune, or did they just relabel "the pox"? Medical knowledge of syphilis and gonorrhea was set back in the eighteenth century when the highly respected biologist John Hunter inoculated himself with what he believed to be a pure culture of syphilis, but which was also contaminated with gonorrhea. From the symptoms he developed, he concluded that the two diseases were caused by the same agent. This error was not finally corrected until the French biologist Philippe Ricord proved the distinct nature of the diseases in 1837. The next important development did not occur until 1905 when Fritz Schaudinn first viewed the syphilis spirochete under a microscope. In the following year August von Wassermann developed a test for syphilis, and in 1909 Paul Ehrlich finally developed a treatment with Salvarsan to replace the horrors and uncertainties of the mercury treatment, but even Salvarsan had a highly toxic arsenic base. And Salvarsan, like mercury, only removed the symptoms of the first stages and did not always prevent the subsequent lethal symptoms of the tertiary stage. So throughout the nineteenth century Europeans had no way of knowing for certain if they had contracted syphilis, they had no reliable treatment for it, and even after they were supposedly "cured" there was no way to ascertain whether the disease would break out at some later stage in life or attack future generations. Under such conditions fear of this particular disease was a source of torment to all who dared to have intercourse with anyone they did not know to be chaste. There was one very good justification for Victorian sexual morality: it was the only reliable prophylaxis against veneral disease.

The fear of syphilis in particular was so acute that it generated an independent derivative disease, a special kind of hypochondria called "syphilophobia." People suffering from this disorder manifested a variety of psychosomatic symptoms of real syphilis and generally suffered from the obsessive thought that they had contracted it, no matter how thorough their precautions. And since the medical theories on the causes of venereal diseases included infection from contaminated towels, beds, drinking glasses, pipes, toothbrushes, razors, pencils, musical instruments, tattooing and kissing—even kissing the Bible—the potential syphilophobe had no end of precautions to make to feed his obsessive fears.

The nature of the disease itself contributed to its mystery and its frightfulness. Syphilis was a disease ideally suited to bear out the admonitions of the zealous Christians who insisted that it was the divinely conceived wages of sin. It was an ideal Protestant disease, as well as an ironically Victorian disease. One transgression, a single sexual contact, could lead to a lifetime of suffering. There was no way of knowing for certain if one had been contaminated or not, parallel to the Christian notion that man cannot ever be certain if he is destined to be saved or not. One was never certain of the cure, and of course none were deserving of cure. No precautions against it were sufficient, paralleling the Christian notion that no human works could possibly influence divine salvation. The diseased were truly condemned to "sickness unto death" and even beyond, through the infection of future generations. Syphilis was known to attack fortuitously, and often the most blatant libertine contracted no symptoms while one-time transgressors caught a heavy dose. Hence there was no way of understanding why some were condemned to suffer while others were not. The ways of the disease, like the ways of God, were seemingly beyond the comprehension of man and subject to the fortunes of fate. The disease also had a mysterious and tormenting timing. After an initial contact a few superficial symptoms appeared and then disappeared spontaneously. After about eight weeks the secondary symptoms appeared and lasted for around nine months. Then the disease entered a period of latency, from which it erupted five to fifteen years later with full force, attacking the vital centers—particularly the heart and nervous system. It was only in the late nineteenth century that physicians succeeded in ascertaining that these later symptoms were a direct consequence of the syphilis spirochete, and not some general punishment for a life of debauchery, as was believed earlier in the century. Hence the progress of knowledge about the mechanisms of infection and transmission of venereal disease contributed at first to the fear of them. This is one instance where the progress of scientific knowledge intensified fears about sexuality and bodily processes, at least until that knowledge had progressed sufficiently to find a complete and lasting cure. The Victorians never found one; for that, the world would have to wait until 1943, when penicillin was employed to eradicate all traces of the disease in an infected person.

The standard procedure for treating syphilis in the nineteenth century was to introduce mercury into the body by any of several methods. Oral introduction of it was achieved by a variety of means: Keyser's pills, Lagneau's lozenges, Olivier's biscuits; even some "mercurial cigars" are listed in one account.[14] For those who could not tolerate oral means, injections or suppositories were employed. For the most sensitive, mercurial ointments were prepared for rubbing into the mucous membrane of the mouth or rectum, or through the skin—usually the groin.

The treatment was almost as frightening as the disease itself. Mercury is effective in raising body temperature and fighting the pathogenic effect of the syphilis spirochete, but it is a poison and can produce a number of toxic side effects. After about two weeks of this treatment the gums become sore and salivation increases sharply. William Acton explained in his treatise of 1857 that salivation was a test of the effectiveness of the treatment, "and spitting, to the extent of half a pint to a pint daily," was to be expected.[15] Imagine the degradation and humiliation of a Victorian man or woman forced to submit to a treatment for syphilis that required them to take a drug which made them spit up nearly a pint daily and which turned their gums black, tainted their breath with a "coppery" odor, loosened their teeth, produced nausea and dizziness—and that more than likely would have no certain positive effect on the disease. To make matters worse, these "cures" were frequently administered by quacks, because the patients were ashamed to acknowledge even to the family physician that they had contracted venereal disease. A final method of treatment was, according to a popular supersition, intercourse with a virgin.[16] The absurdity and tragedy of this bit of medical witchcraft shows the extent to which the panic of infection led Victorians in their search for relief.

The general fear that the incidence of venereal disease was increasing led to some efforts at social control toward the end of the nineteenth century. The most notorious of these were the Contagious Disease Acts passed in England in the 1860s. The first act, in 1864, provided for medical examinations of any woman accused of having communicated venereal disease to military personnel. An Amending Act of 1866 allowed for periodic medical examination of prostitutes by magistrates. The surgeon's certificate of contamination was sufficient to detain her in a hospital for up to three months. In 1869 the act was further amended so that any person could give information under oath before a magistrate that a woman was a prostitute, whereupon a notice was to be served upon her for a compulsory examination. She might then be detained for up to nine months if necessary. In the same year Josephine Butler organized the "Ladies' National Association for the Repeal of the Contagious Disease Acts," which inaugurated a campaign that eventually got the Acts repealed in 1886. Though the acts were unabashedly biased against women, they represented the recognition that contagion is a problem of human existence that demands some form of societal control. A more effective response to the problem of venereal disease had to await the movements that emanated from the First International Congress on Venereal Disease, which met in Brussels in 1902.

Chapter 5

*Olfactory Ontology
and Scented Harmonies*

> What separates two people most profoundly is a different sense and degree of cleanliness. What avails all decency and mutual usefulness and good will toward each other—in the end the fact remains: "They can't stand each other's smell!"
>
> Nietzsche—Beyond Good and Evil

Society's willingness to recognize the role of smell in human affairs is a good index of its willingness to recognize that human beings are corporeal beings, closely linked to the animal world. The hairiness of man and his smells are constant reminders of his animal ancestry. The Victorians did not handle that side of human existence with ease, though the latter decades of the nineteenth century revealed a good deal of research on the role of body odors and the sense of smell in human affairs. That research is the most convincing proof that after 1850 European society began to come to terms with the corporeal side of human existence and to challenge the repressive sexual morality that so strongly influenced European life in the early part of the century.

In spite of the emphasis on the role of the senses in eighteenth-century psychology, the function of the "lower senses"—

taste and smell—was largely neglected by philosophers and psychologists alike. Kant gave little importance to the sense of smell, and there were few studies of it in the early nineteenth century.[1] Hippolyte Cloquet's *Osphrésiologie, ou traité des odeurs, du sens et des organes de l'olfaction* (1821) offered the first systematic study of the physiology and psychology of the olfactory system. Interest in smells was revived around mid-century because of the popular theory that odors emanating from putrefying organic bodies caused disease—particularly cholera, which had erupted in epidemics in 1831 and again in 1848. The desire to eliminate miasmatic origins of disease became linked with the campaign to provide adequate ventilation in hospitals, factories, schools, and homes. In the summer of 1858 the level of the Thames sank sharply from lack of rain and created the "Great Stench" that choked London. The stink was so nauseating that cloths soaked in chloride of lime were hung over the windows and doors of Parliament to neutralize the smell during sessions.[2] The English epidemiologist William Budd commented on the historical significance of the Great Stench: "Never before had a stink risen to the height of an historic event."[3] The universal discomfort that it caused resulted in public support of Florence Nightingale's efforts to improve ventilation in hospitals.

As the public health movement began to get under way around mid-century, and as life became cleaner, Europeans began to become more sensitive to smells, both to body odor and to foul air. The possibility of keeping streets and bathrooms, as well as one's body, clean led to a growing sensitivity to smells, a sensitivity that we see developing in the latter decades of the century. Before that time, I assume, body odor was so overpowering that heightened sensitivity to smell was discouraged rather than cultivated.[4]

In 1871 Darwin's book *The Descent of Man* stirred interest in the role of the olfactory system in sexual life. Following his lead a number of studies attempted to argue that the sense of smell, which played an essential part in the sex life of animals, was still operative in human sexuality, even though society had minimized its importance. In 1886 the Frenchman Auguste Galopin expanded Darwin's speculation about the central role of odors in human love. In his classic *Le Parfum de la femme* he argued that the mutual interaction of odors constitutes the essence of sexual love. "The purest marriage that can be contracted between a man and a woman is that engendered by olfaction and sanctioned by a common assimilation in the brain of the animated molecules due to the secretion and evaporation of the two bodies in contact and sympathy." [5] Galopin's insistence that love is smell defied romantic conceptions of the spiritual nature of love by insisting that the sniffing about so long associated with the sexual life of animals was also a central element in the love life of humans.

In 1891 the German embryologist Ernst Haeckel in his popular textbook *Anthropogenie* elaborated on Darwin's theories about the evolutionary history of smell and its relation to sex. Haeckel reasoned that sexual attraction must have originally arisen through smell or taste, an "erotic chemotropism" that operated through water to bring two cells together. These first sex cells must have had some kind of basic mental activity which was linked with their sense of smell. This primeval link between smell and sex continued to exist in the racial memory and remained related to sexual stimulation. Smell as a human sexual stimulus was a vestige from the age when it alone received sexual sensations and directed the cells toward one another. Haeckel's theory was widely accepted around the turn of the century.[6]

Freud's intimate confidant during the 1890s, Wilhelm Fliess, devoted a large part of his gynecological research to establishing the connection between nasal and sexual processes. He observed that during menstruation the capillaries of the nose swell and that application of cocaine to special "genital spots" of the nose reduced menstrual pain.[7]

The most complete study of the connection between smell and the sexual life of man was published in 1900 by a German physician, Iwan Bloch.[8] After surveying the history of the study of smells, reviewing some of the essentials of the physiology of the olfactory system, and recounting the most recent efforts to link sex and smell, Bloch offered his own analysis of the significance of smell in modern life. He regretted the contemporary cultural neglect of the sense of smell in human affairs, particularly in relation to sex life, but he was equally disturbed by exaggerations of the importance of smell and certain perversions of it that he discovered in the work of Baudelaire, Zola, and Huysmans. He concluded by calling for an effort to restore the pleasure of natural smells in order to maximize the enjoyment of life.

Bloch's study contained traditional folklore on the interconnection of the nose, the sense of smell, and sex. The idea that a large nose betokened large genitals and sexual potency was part of popular culture in Europe over the centuries. He also discussed the old belief that having intercourse with or smelling young people will rejuvenate the sexual potency of an older person.[9] One man, banking on this theory, slept with two young wet nurses and took milk from both of them. Bloch then recounted some attempts to make bad breath grounds for divorce. Paul Mantegazza was his authority that "stinking breath" was grounds for divorce in various law books, although Bloch did not list the books himself. He added from his own research that the Leipzig medical faculty seriously discussed whether bad breath was sufficient grounds for divorce, and he concluded with his own supposition that many unhappy marriages may be explained by "antipathetic smells." His discussion of bad breath as a possible cause of marital difficulty was a sober reminder of the intrusiveness of the body in human affairs. That the relationship between a married couple just might fail because of bad breath or body odor was a sharp contradiction to the rhapsodic eulogies about lofty ideals for marital happiness appearing in the countless *feuilletons* and family magazines that were so popular throughout the Victorian age. Bloch's sources suggest that Nietzsche's remarks, quoted at the beginning of this chapter, might have applied to a surprisingly large number of married couples.

Acceptance of the sexual role of smell led the Victorians to further question their constricting prudery. Havelock Ellis, the most prolific and outspoken critic of Victorian prudery, devoted almost eighty pages of his monumental *Studies in the Psychology of Sex* to the role of smell. Ellis's study of smell is largely a bibliographical essay which surveys the literature topic by topic—the history of the study of smell, the psychology of smell, smell in literature, and the perfume industry. His own comment on the fate of the olfactory system in civilized society avoided the equivocation of Bloch. Ellis lamented that the sense of smell had given way to the sense of vision which almost exclusively dominated modern sex life. He remained steadfastly convinced that "the latent possibilities of sexual allurement by olfaction, which are inevitably embodied in the nervous structure we have inherited from our animal ancestry, still remain ready to be called into play."[10]

The idea that modern civilization had tended to deny and inhibit the role of smells in human existence—sexual and other—was suggested by a number of scholars by the turn of the century. Though their evaluation of that particular repression of human sensual existence varied, all generally agreed that something had been lost. Freud concurred, but broadened his interpretation by relating denial of smell to denial of human corporeality in general. He further argued that the repression of smell imposed by civilization was a major cause of mental illness. "With the assumption of an erect posture by man and with the depreciation of his sense of smell, it was not only his anal eroticism which threatened to fall a victim to organic repression, but the whole of his sexuality." [11]

The famous experts on sexual pathology in the late nineteenth century—Krafft-Ebing and Bloch in Germany, Charles Féré and Alfred Binet in France—agreed that a large number of prevalent fetishes involved smells. The popularity of the handkerchief, the shoe, and underclothing, as well as feet, sweat, and excrement, for such purposes was partly due to the pungent bodily odors associated with them. Bloch explained shoe fetishism by the similarity between the smell of shoe leather and the female genitals. He agreed with Krafft-Ebing's explanation that the rise of the incidence of smell-related fetishes was an exaggerated counter-reaction to general cultural suppression of the sense of smell.

The scientific study of smells was by no means limited to the relation between smell and sexuality. Other studies involved the role of smell in stimulating memories, the ethnology of smells (so-called racial or national smells), and the general psychology of smell.

The famous English physician Henry Maudsley touched on an aspect of the psychology of smell that received growing attention in the later decades of the century—the ability of smell to trigger memories. In *The Physiology and Pathology of the Mind* (1867) he argued that no sense has so strong a power for calling up memories with deep "emotional reverberation." In 1896 the French psychologist Théodule Ribot expanded upon Maudsley in a general discussion of the memory of feelings. He argued that tastes and smells have particular ability to revive feelings precisely because they are so difficult to remember. Also they are associated with organic and physiological states and hence are able to generate the physical and emotional changes produced at the time of the original feeling. This psychological speculation about the mnemonic power of smells and tastes became the theoretical foundation for one of the greatest literary works of the early twentieth century: Proust's *Remembrance of Things Past.* In the first volume the narrator, Marcel, stirred by the taste of a *petite madeleine,* speculates that "what is thus palpitating in the depths of my being must be the image, the visual memory which, linked to that taste, has tried to follow it into my conscious mind." Marcel's rumination inspires him into a long search for the lost world conjured up by the taste of the little cakes. In the final volume, *The Past Recaptured*, the taste of a *petite madeleine* brings back the feelings and sensations of the world of his childhood. In his exploration of a world that was unlocked by a combination of tastes, Proust underlined the extent to which our entire emotional life is linked with that part of our sensorium dismissed for so long by psychologists as the "lower senses."

The most zealous argument for the importance of smells in human existence was made by the German biologist Gustav Jaeger in a series of essays published in 1881 on "the origin of the soul." [12] He began with the complaint that modern man has lost his sense of smell. This sense, he said, plays a major role in man's most intimate and vital functions and is essential to his sensuous relation to the environment. He elaborated upon Schiller's often-repeated quotation, "While philosophers dispute, hunger and love decide our fate," by arguing that both hunger and love are chemical processes that rely on smell. The intimate relation between a mother and a child is enhanced by odor, and some children detect their mother's smell and will not nurse from a strange breast. The family bond is revealed by the harmonious mixture of the smells of the mother and her children. Different races have unique smells, and various emotional states (such as terror and hatred) produce specific smells. Jaeger tested this theory on his daughter by smelling her hairnet after she exercised to prove that she smelled different in calm and excited states. The startling conclusion of all these ruminations is that odor is the "origin of the soul."

Jaeger achieved more recognition in Europe and America for his system of woolen underwear than for his exaltation of the olfactory system, but the two interests were related, as he explained in *Health-Culture and the Sanitary Woolen System*. Jaeger insisted that everybody ought to wear a complete set of woolen underwear at all times to help keep the skin uniformly warm, to offer an outlet for perspiration, and to stimulate the skin and circulation. The arguments were not unreasonable in an age when women often wore no underwear and were subject to extreme differences in body temperature, and when clothing often did not allow for adequate perspiration; but Jaeger added to these purely hygienic considerations his strange theory of the importance of odors in regulating physical as well as emotional life. "Disease is stink" and, he argued, the "offensive effluvia" emanating from privies are dangerous when they contain germs. Since the proper perspiration of the skin permits the body to expel these disease-causing odors, he claimed that woolen underwear would maximize health by reducing the "noxious principle within the body." [13]

In 1903 a French biologist published a catalog of various kinds of body odors, normal and pathological.[14] His attempt to characterize such specific odors as those of armpits, breath, feet, and genitals reads nowadays like a satire on nineteenth-century quasi-scientific scholarship. A more complete catalog was published by the Dutch psychologist H. Zwaardemaker, whose *Physiology of Smells* (1895) outlined nine specific categories. In 1887 Zwaardemaker developed the olfactometer—an instrument designed to put the study of smells on a scientific basis by providing a means of measuring the threshold of the sense of smell. His work was a major effort to surmount the problem of subjectivity which had made the experimental study of smells impossible. In 1894 the German physiologist Carl Giessler suggested that smells could be divided into two major categories—idealizing and unidealizing—the division being determined by the effect the respective odors had on human digestive and sexual processes. Like many of his predecessors Giessler concluded with the hope that the sense of smell would be further cultivated.[15] In the following year Heinrich Pudor joined the campaign for the revival of the sense of smell, lamenting the general anesthetization of the human senses that modern life appeared to be producing.[16] Pudor was soon to turn his energies to support physical culture, and in particular the movement for nudism and *Freikörperkultur*.

The complaint that the sense of smell had been neglected

was not limited to the obscure psychologists and physiologists we have so far surveyed. Three important French literary figures—Baudelaire, Zola, and Huysmans—began to write graphic accounts of the smells of everyday life.

Baudelaire used the associations with smell to enhance the imagery of his poetry. He knew Théophile Thoré, who in 1836 published *Art des parfums*, which developed the idea that smells can be as expressive as colors. Thoré argued that while painting and sculpture represent the object directly, perfumes reveal the intuition of things, like music.[17] Baudelaire explores the artistic possibilities of this theory in his famous poem *Correspondances* (1857):

> There are perfumes as fresh as the flesh of a child
> Sweet as the oboes, green as the meadow;
> And others corrupt, triumphant and rich,
> Opening out like infinite things,
> Like amber, like musk, benjamin, incense
> Singing the raptures of spirit and senses.[18]

Baudelaire thus relates smells to touch (child's flesh), sound (oboes), vision (green as the meadows), and emotional states (corruption and triumph). He exalts woman's body odors, particularly the smell of her hair, in several poems in *The Flowers of Evil*. *Parfum exotique* describes the sensuous pleasure of the smell of a woman whose breath is a guide "toward charming climates." In *La Chevelure* he longs to lose himself in the smell of a woman's hair:

> O aromatic forest!
> As other spirits travel on music
> Mine, O my love! Swim upon thy scent.[19]

He dwells further on the sensuous pleasure of a woman's hair in *Un Hémisphère dans une chevelure* (1869). "Let me breathe forever the scent of your hair and immerse my whole face into it. . . . What things I hear in your hair! My soul travels on fragrance as the soul of other men travels on music. [In your hair] the air is perfumed by fruit, by leaves, and by human skin. . . . In the burning hearth of your hair I breathe the scent of tobacco mixed with opium and sugar." [20]

Zola's fascination with odors was sufficiently pronounced in his work to inspire a complete study by Léopold Bernard, a Professor of Philosophy at Montpellier. Bernard's essay credited Zola with elevating the impoverished language of odors by including descriptions of the fine nuances of smells in his novels. Zola's materialism, Bernard argued, convinced him that man is essentially a system of organs, and those involved in nutrition and reproduction are naturally dependent on the olfactory system. Each body has its special odor, and the novelist ought to try to describe it in order to reveal the character he wishes to portray. "For Zola, the description of a character is not complete unless he has noted the expressive odor which he exhales." Interest in Zola's preoccupation with odors led to the performance of an autopsy of his olfactory system. The examination showed that its development was "below normal." [21]

Bernard surveyed those novels in which attention to smells is particularly obvious. In *La Faute de l'abbé Mouret* (1875) Albine is described as smelling like a bunch of flowers, the effeminate Serge has been "robbed of his manly smells," and the virgin Désirée smells like health. When Albine commits suicide because her love with Mouret is impossible, she fills her room with bouquets of flowers and is asphyxiated by their fragrance. *Le Ventre de Paris* (The Belly of Paris) includes detailed accounts of the odors of the marketplace, and in *La Curée* Zola explores the smells of lovers and points out similarities between the odors of religion and sex. In *Nana* (1880) Zola fills out his description of Count Muffat's first visit to see Nana backstage at the theater with an account of its "haunting" smells: a composite of "the odor of gas, the glue used in the scenery, the dirt in dark corners, the underwear of the chorus-girls . . . the acidity of the toilet waters, the perfumes of soaps, the putrid smells of exhalations . . . an odor of womanhood—a musky scent of make-up mixed with the primitive smell of human hair." [22] The most famous instance of Zola's account of smells—the symphony of cheeses in *L'Assommoir*—no doubt was the inspiration for a similar "symphony of odors" conducted by the hero of Huysmans's *Against Nature*.

While Zola kept his descriptions of smells in language that a layman could easily understand, Huysmans gilded his accounts with the specialized language of the world of medicine and chemistry. In a series of sketches of Parisian life published in 1880, Huysmans included a short essay, *"Le Gousset,"* on the varying odors of women's armpits.[23] He reported having followed a group of woman gleaners working in the sun and discovered that their body odor had an alkaline sting. "It seized you, irritating your mucous membrane with a rough odor which had in it something of the relish of wild duck cooked with olives and the sharp odor of the shallot." Huysmans spoke as a connoisseur when describing the mixture of perfume and body odor on women at fashionable ballroom parties. "There the aroma is of ammoniated valerian, of chlorinated urine, brutally accentuated sometimes, even with a slight scent of prussic acid about it, a faint whiff of overripe peaches." These "spice-boxes," he continued, become especially seductive when their body odor is filtered through garments heavy with an accumulation of perspiration. Huysmans also described body odor as varying with the color of the skin and hair. "Audacious and sometimes fatiguing in the brunette and the black woman, sharp and fierce in the red woman, the armpit is heady as some sugared wines in the blondes." He concluded that the smell of armpits covered "the whole keyboard of odors."

In his next novel, Huysmans developed that metaphor into an imaginary instrument on which to play symphonies of smells. The aristocratic hero of *Against Nature* (1884), the dissolute Des Esseintes, retreats from the banality of bourgeois life to a country estate where he begins to lead a life of "studious decrepitude" which involves a systematic indulgence and exploration of each of his senses. He lavishes upon himself combinations of strange colors and sounds, unnerving readings, and exotic art. He devises a "mouth organ"—a modified organ that tips forward a small vial of different liqueurs as he pushes on different keys—in an effort to play "internal symphonies" to himself with liqueurs representing musical instruments. And so he composes quartets and symphonies orchestrated by oboes (kummel), trumpets (kirsch), and coronets (gin). Des Esseintes explores the possibilities of olfactory sensation with the exactitude of a scholar. He seeks to master "the grammar, the syntax of smells," and to come to understand the rules that govern them. He ventures to develop "scented harmonies" and "aromatic stanzas" by opening stoppered bottles containing various odors in specified combinations and sequences. He attempts to "unscrew the separate pieces forming the structure of a composite odor" and begins to develop to a high degree his ability to detect and enjoy different smells.

Against Nature was intended to assail the popular idea that natural experiences and pleasures represented the height of human sensual enjoyment. Huysmans believed that nature was boring, and that only artificial sensations produced by the inventive human mind are worthy of aesthetic interest. His exploration of the self-indulgent sensualism of Des Esseintes struck many of his readers as the height of perversity, and the novel quickly became a manifesto for the Decadent Movement in literature and art. Its reverent attention to smells was a challenge to drawing-room sensibilities and a further reminder to a generation in flight from the animal and corporeal origins of life that human existence is not merely cerebral and spiritual, but has a persistent material basis in the human body.

Jaeger's wild olfactory ontology, Zola's symphony of cheeses and Huysmans's "scented harmonies" were minor side shows in the general course of European culture. But they were also symbols of protest against the tradition of denial and neglect that had for so long suppressed serious consideration of the role of smell in human affairs and that had so persistently viewed it as the "lowest" of the human senses.

Chapter 6

Sexual Alienation and Sexual Selection

Both Marx and Darwin made substantial contributions to the understanding of the corporeal determinants of human existence. In their respective works they argued that history and evolution are in the service of bodily needs. Each historical period, like each evolutionary stage, may satisfy those needs differently according to circumstance, but those needs are the mainspring for human activity, and any science of man must begin by understanding them.

In *The German Ideology* Marx's first premise of his theory of history is the elementary proposition "that men must be in a position to live in order to be able to make history." To this hypothesis he made a marginal note that included a list of three topics: human bodies, needs, and labor.[1] Those topics, I believe, outline the essentials of his argument from biological necessity to the history and social life of man. Investigation of human existence must begin with the recognition that certain bodily needs determine man's vital activities—production of food and housing and reproduction of the species.

The final topic listed—labor—interested Marx far more than the biological needs first mentioned. His social theory is largely an account of the way man "works up" his environment to satisfy his physical needs. Historical analysis ought to contain an account of the dialectical interaction between the material conditions in which man labors to satisfy his bodily needs and the institutions and ideas created during that laboring process. Although Marx was particularly challenged to explain the interactions between the substructure and superstructure of successive historical periods, he recognized that fixed biological needs supplied the incentive for all human effort. His analyses of various historical forms engaged him in explaining ideas and institutions, but the starting point of all his analyses, at least the *logical* starting point, would always be bodily needs and laboring man. Marx's insistence on the corporeal determinants of human existence challenged the idealist tradition that had previously dominated the writing of history.

Ludwig Feuerbach's philosophy marked an important step in the rejection of Hegelian idealism. Marx embraced it before he eventually dismissed it as "contemplative materialism," but it played a decisive critical role. In 1841 Feuerbach challenged idealism with *The Essence of Christianity*. Engels has vividly recorded its impact: "Without circumlocution it placed materialism on the throne again. Nature exists independently of all philosophy. It is the foundation upon which we human beings, ourselves products of nature, have grown up. Nothing exists outside nature and man." [2] Engels continued: "One must himself have experienced the liberating effect of this book to get an idea of it. Enthusiasm was general; we all became at once Feuerbachians." Feuerbach had argued that the "essence of Christianity" was a series of idealized projections of human nature into an imaginary construct—God was a collection of anthropomorphisms. Another source of enthusiasm for the book was its deification of love in place of reason as the guiding force in human history. In 1843 Feuerbach further emphasized the sensuous element in his new philosophy: "The old philosophy had its point of departure in the proposition: I am an abstract, a merely thinking being to which the body does not belong. The new philosophy proceeds from the principle: I am a real and sensuous being; indeed, the whole of my body is my ego, my being itself." [3]

When Marx began to work out his philosophy in 1844 he was strongly under the influence of Feuerbach. His explanation of the nature of alienation was also a major concern of the romantic movement in Germany, which was striving to break the grip of asceticism over German emotional life and sexuality.[4] Marx's *Economic and Philosophic Manuscripts* of 1844 repeatedly insisted on the sensuous basis of human existence. Man is part of nature, and "that nature is his body with which he must remain in continuous intercourse in order not to die." [5] Marx then spelled out the sensuous nature of man's encounter with the world: "seeing, hearing, smelling, tasting, touching, thinking, observing, feeling, desiring, acting, loving—in short, all the organs of his individuality . . . are . . . the appropriation of human reality." [6] His repeated arguments for the sensuous nature of man's interaction with the world are forceful polemics directed against the metaphysical puff that filled so much German philosophy. "To say that man is a corporeal, living, real, sensuous, objective being full of natural vigor is to say that he has real, sensuous objects as the objects of his being or of his life, or that he can only express his life in real, sensuous objects." [7] When Marx turned to consider human interaction, the general sensuousness became explicitly sexual. "The immediate, natural, and necessary relation of human being to human being is also the *relation of man to woman*. . . . In this relation is *sensuously revealed* . . . the extent to which human nature has become nature for man. . . . From this relationship man's whole level of development can be assessed." [8] His assessment of capitalistic society would focus on the fate of this relationship as symptomatic of a general process of alienation that estranges man from nature, from his own body, and from gratifying sexual relations. Bourgeois marriage and prostitution, he believed, were the institutions of alienated sexuality.

Marx also shared Feuerbach's contempt for the asceticism so firmly ingrained in the German mind from the solid union of Christianity and the bourgeois-capitalistic work ethic. In a famous passage he railed against the self-denial inherent in both ethical doctrines. "The less you eat, drink, buy books, go to the theatre or to balls, or to the public house, and the less you think, love, theorize, sing, paint, fence, etc., the more you will be able to save and the *greater* will become your treasure which neither moth nor rust will corrupt—your *capital*. The less you *are*, the less you express your life, the more you *have*, the greater is your *alienated* life and the greater is the saving of your alienated being." [9] The central idea of this passage—that the more one denies oneself, the more one makes of oneself—reveals a perplexing contradiction that lies at the heart of Western asceticism. Some of the most ambitious thinkers of the past century have addressed themselves to it. Marx viewed it largely as a consequence of the particular psychological effects of capitalism, particularly the drive to accumulate more and more profits by not spending money on oneself and by competing with others for the material goods of the world. Nietzsche, as we shall see, attempted to explain asceticism as a necessary stage in the development of the human mind, and one which he believed he could illustrate by tracing the development of the Judeo-Christian morality. Max Weber believed that asceticism was essential to the development of capitalism. Freud came to be interested in the consequences of sexual asceticism in the genesis of mental illness. He viewed self-denial and instinctual repression as a necessary condition for, and consequence of, the progress of civilization. All of these thinkers agreed, however grudgingly, that great progress was made possible by self-denial, thrift, or instinctual repression. They debated about the price man had to pay for these accomplishments, but they were all duly impressed by them.

Though Marx's appreciation of the sensuous underpinnings of human existence remained constant throughout his life, the intoxication with Feuerbach's thought lasted only a few years, and in the spring of 1845 Marx began to work out a critique, subsequently published as *Theses on Feuerbach*.

Marx credits Feuerbach for attacking mere "abstract thinking" but claims that he ultimately failed to develop a materialism that viewed social reality as something that develops out of concrete historical situations and, as Marx believed, that must be changed by revolutionary actions. Feuerbach employed what Marx termed "sensuous contemplation," which involved thinking about the sensuous basis of life, but as abstracted from specific historical situations. Moreover, Feuerbach's philosophy did not seek to understand historical change as part of reality. Rather, he searched for fixed essences—of human nature, of religion, of social organization. Marx makes his objection explicit in the first of the "Theses": "The chief defect of all hitherto existing materialism—that of Feuerbach included—is that the thing, reality, sensuousness, is conceived only in the form of the object or of contemplation. . . . [Feuerbach] does not grasp the significance of 'revolutionary,' of 'practical-critical,' activity."

Marx's first detailed formulation of his philosophy of history in *The German Ideology* (1845-6) emphasized the biological basis of human existence and hence of history. In historical inquiry "the first fact to be established is the physical organization of these individuals and their consequent relation to the rest of human nature." [10] His study of history will begin with "men in the flesh." "We set out from real, active men, and on the basis of their real life-process we demonstrate the development of their ideological reflexes and echoes of this life-process. The phantoms formed in the human brain are also necessarily sublimates of their material life-process." [11] The study of man and society, the history of political institutions, the history of art and religion, all derive from man's physical organization, specifically the biological drives which require that he labor to survive. Marx's philosophy, in contrast with German philosophy "which descends from heaven to earth," begins on earth and ascends to man's higher cultural achievements. Therefore, each period is characterized by the nature of man's relation to the productive forces in his society.

Marx nowhere renounced his original insistence on the physical basis of life and of history, but because the biological endowment is constant, he devoted his attention largely to labor and to the forms of social organization that derive from it. His interest in labor was torn between contrasting views. On the one hand he revered labor as the only source of value—spiritual as well as material. On the other hand he viewed alienated labor as the curse of modern man. In the capitalist age in particular, he held, labor had arrived at its most intense form of alienation. The task of reason was to restore to the laborers control over, and enjoyment of, their own creative powers.

Marx's definition of labor in *Das Kapital* emphasized both its own physical nature and the mediating role it plays in the reciprocal interchange of men and nature. "Labor is, in the first place, a process in which both man and Nature participate, and in which man of his own accord starts, regulates, and controls the material re-action between himself and nature. He opposes himself to Nature as one of her own forces, setting in motion arms and legs, head and hands, the natural forces of his body, in order to appropriate Nature's productions in a form adapted to his own wants. By thus acting on the external world and changing it, he at the same time changes his own nature." [12] Against this sketch of the natural condition of the laborer, Marx contrasts the situation of the laborer under capitalism, where the laborer loses his autonomy and is *not* able to labor "of his own accord," but rather sells his labor for wages. Moreover the entire labor process is "alienated" in a number of ways. The wage-laborer does not affirm, but denies himself and does not develop his mental and physical energy. The more impressive the machinery at which he works, the more powerless the laborer becomes. He no longer is able freely to "work up" the materials of nature but is forced to work out of necessity like an animal. The labor becomes so repetitious that it alienates man from his own body. In the factory, man becomes a "mere living appendage" to machines. Marx's critique of the problems facing the laborer under capitalism emphasizes the effect of alienated labor on the human mind and the human body. The essential activities of labor—the motion of "arms and legs, head and hands, the natural forces of his body"—cannot be freely executed under capitalism, which robs labor of its creative qualities. The robotlike appendage of the factory machine, satirized by Chaplin in *Modern Times*, becomes the new human biological product of the capitalist mode of production. Just as the formation of the five senses was the work of all previous history, as Marx saw it, so the human body will be further modified by its relation to machines and the factory system. The harmonious and sensuous interchange of man and nature for the purpose of satisfying bodily needs is destroyed by the competition that divides man from man, by the machines that cut up the work process, by the system of private property that separates the worker from the means of production, and by the fetishistic overvaluation of commodities that leads to an obscuring and devaluation of human needs. Each of these nefarious effects of capitalism Marx described with metaphors that suggested division or separation, and one consistent casualty of the capitalist system is the relation of man to his own body. This concern of Marx is amply demonstrated in *Das Kapital*, in which he elucidated the disastrous effects of various occupations on the human body. In addition to his catalog of the many diseases of the eyes, the skin, and the lungs, he evaluated the ruinous effect that the factory system has on the nature of the labor process and on the entire human body. Marx offered a clear reminder that human existence derives from, and remains intimately tied to, its biological underpinnings.[13]

On the last page of *The Origin of Species* Darwin wrote what proved to be the most important prediction in the history of the social sciences. Concluding his monumental study of worms and bugs, pigeons and fish, he wrote a single sentence on the possible consequences for the study of human life that his work might have: "Much light will be thrown on the origin of man and his history." By the time he published his own application of the theory of natural selection to the origin of man, a score of other scientists had already made the attempt. *The Descent of Man* (1871) summarized the changing conception of human nature that had been developing since the publication of his classic twelve years earlier.

Though the wealth of data he adduced to prove the theory of evolution often tended to obscure the general argument, Darwin was able to strike his readers with shocking formulations of the radical changes his biological theories implied. If the lazy reader did not quite follow what all the discussion about embryos had to do with the evolution of man, Darwin summarized the message with the stunning conclusion that "man is descended from a hairy, tailed quadruped, probably aboreal [sic] in its habits."[14] That formulation of his theory raised a storm of protest that abated only as Europeans came to accept Darwin's new view of human nature. Darwin's theory made four revolutionary changes in the going conception of human nature at that time. The theory of the animal origins of man contradicted the idea of a special creation of man; the explanation of animal rudiments demolished the idea of man's having been fashioned in God's image; the theory of sexual selection introduced in *Descent of Man* overturned the romantic conception of human love and emphasized the animal nature of human sexual relations; and his philosophical materialism brought the human mind into nature and viewed it as dependent upon the body.

To substantiate his theory that man had descended from animal forebears, Darwin surveyed a number of remnants of lower forms that persist in the human body. The general bodily structure of man, Darwin argued, resembled that of various animals. "All the bones in his skeleton can be compared with corresponding bones in a monkey, bat, or seal." The fact that man can contract certain diseases from animals "proves the close similarity of their tissues and blood." The embryological development follows the same pattern as that of some lower animals. He quoted from the German embryologist Von Bear that "the feet of lizards and mammals, the wings and feet of birds, no less than the hands and feet of man, all arise from the same fundamental form." [15] The most disturbing evidence Darwin presented to show the animal origins of man was no doubt his discussion of the various rudiments that persist in the human body. The muscles for twitching the skin or moving the ear, the sense of smell, wisdom teeth, male mammaries and the appendix are survivals of this nature. "It is, as I can now see, probable that all organic beings, including man, possess peculiarities of structure, which neither are now nor were formerly of any service to them, and which, therefore, are of no physiological importance." [16] The idea that we carry a load of obsolete biological baggage which is of no use to us further demolished traditional views of human nature and ridiculed the notion that man had been created in God's image.

Darwin did not rest his case solely on evidence of the biological similarities between animals and men; he extended his argument to include comparisons of intellectual and emotional similarities. His object, he wrote, was "to show that there is no fundamental difference between man and the higher mammals in their mental faculties." [17] And so he relentlessly argued that emotions such as terror, courage, timidity, and love are clearly manifested in certain animals. Even the aesthetic sensibilities of man resemble those of animals, and in one passage Darwin argued that the aesthetic sensibilities of birds might be more highly developed than that of savages who admire "hideous ornaments" and "hideous music."

If the fact of human evolution from animals was a "blow" to the Victorian ego, the mechanisms of that evolution were further insulting, at least to those who thought that the survival of the fittest or sexual selection was inhumane and bestial. In *The Origin of Species* Darwin had emphasized the role of the struggle for existence and the survival of the fittest. "Thus, from the war of nature, from famine and death . . . the production of the higher animals directly follows." In *Descent of Man* he introduced another mechanism—sexual selection—to explain the survival of certain traits such as the peacock's feathers, which make the peacock most vulnerable to annihilation in the struggle for existence, but which work favorably in the battle for a sexual partner. He concluded that many bodily structures and mental faculties have been selected because they render individuals more effective in the struggle to find a mate. This addition to his theory further altered the older conception of human nature held by a generation which was as shocked by the idea of sexual determinants of human existence as by its animal origins.

Besides the many physical differences between men and women, Darwin listed some of the mental traits he believed had been evolved by sexual selection. His speculations about male and female character, as we shall see in a subsequent chapter, were uncritically accepted by many people well into the twentieth century. Darwin argued that men are more courageous, pugnacious, and energetic, while women have greater tenderness, intuition, and generosity. Men are better at deep thought, reason, and using their hands. All of these traits, he believed, aid the respective sexes in their search for a sexual partner. In one extraordinary passage Darwin speculated about the effect of inbreeding in the creation of class characteristics. "There is, however, reason to believe that in certain civilized and semi-civilized nations sexual selection has effected something in modifying the body frame of some of the members. Many persons are convinced, as it appears to me with justice, that our aristocracy, including under this term all wealthy families in which primogeniture has long prevailed, from having chosen during many generations from all classes the more beautiful women as their wives, have become handsomer, according to the European standard, than the middle classes." [18]

Darwin challenged the traditional mind/body dualism and the special status given to the mind. He argued that the evolution of higher forms came about solely from the action of natural forces, and that the brain itself is a product of this evolutionary process. He was reluctant to expose publicly the full extent of his philosophical materialism, but in his private notebooks he recorded the materialistic implications of his theory. "Experience shows," he wrote, "the problem of the mind cannot be solved by attacking the citadel itself.

... The mind is a function of body." [19] Ernest Becker has evaluated the historical impact of this aspect of Darwin's thought. "Under the influence of Darwinism, late nineteenth-century philosophy took the mind down from its previously privileged stature, and included it within nature. Mind could no longer be thought of as separate from body: mind served body, and since the main preoccupation of organisms is the negotiation of action, mind was seen to subserve action." [20]

The last paragraph of *Descent of Man* takes the reader from one end of the spectrum to the other, from the exaltation to the degradation of mankind. Darwin begins with an observation that those who may think that mankind has been insulted by his theory ought rather to feel pride at having risen to the summit of the organic scale. He then summarizes many of the "noble qualities" man has achieved. But then, in the final sentence, he drives home that part of his theory which shattered the idea of man's special creation in the image of God: ". . . with all these exalted powers—Man still bears in his bodily frame the indelible stamp of his lowly origin." The ethical implications of the word "lowly" could hardly be missed, despite Darwin's insistence that the correct view of the evolution of the species led to a greater admiration for the achievement of higher forms out of the primeval slime.

Though it is not my intention to suggest that the major significance of Marx and Darwin was their demonstration of the corporeal determinants of human existence, nor that this was the only point of juncture between their systems, the similarity of this one aspect of their views lends strong support to the general picture of European culture I am tracing. It is significant that they both repeatedly felt the need to remind their readers that man *was*, or at least ought to be, superior to animals. Darwin came to argue that the struggle for existence could not be taken as a basis for ethics, precisely because men were not mere animals, and Marx passionately insisted that although the life of the industrial worker had become animal-like under capitalism, it was the essence of human existence to transcend the realm of necessity that governs the animal world and enter the realm of free human existence. The fact that they had to remind the reader that men are not animals suggests that they did show that many human activities share an origin in the animal world.

In the following chapter I shall survey some other developments in European materialism which explore the bodily basis of human existence. The work of Marx and Darwin shared one idea that we shall see was developed by a variety of different thinkers: that man is a sensuous creature, forever driven by his bodily needs.

Chapter 7

Materialism and the Mind-Body Problem

While Marx and Darwin were exploring human corporeality as it related to their systems of thought, a number of other cultural figures were considering other aspects of the same subject. This chapter will survey the work of some philosophers, physicians, and psychologists who argued for the material basis of life. The concluding section on the introduction of surgical anesthesia and the psychology of pain provides an opportunity to speculate on another aspect of the growing awareness of human physicality and its possible impact on the historical development of levels of tolerance for cruelty and pain.

Throughout history there have been periods in which a materialist view of human nature has come into vogue, when philosophers argued that human existence is nothing but a function of the organization and motion of material particles. The revival of materialism in modern times dates from the late seventeenth century when Locke's epistemology provided the theoretical foundation for sensationist psychology. In their campaign against Christian asceticism a number of eighteenth-century philosophers championed the hedonistic psychology and ethics implicit in Locke's philosophy. In addition to the ethic of pleasure, a number of philosophers conceived of human existence in purely materialistic terms. The idea of describing men as machines came from

Descartes, who limited his use of the metaphor to animals, whom he described as "mere machines or automata." In the eighteenth century Lamettrie applied the term to man in a book with the shocking title *Man a Machine*, which rejected Descartes's restriction of the term to animals and concluded that "man is a machine, and that in the whole universe there is but a single substance differently modified."[1] The human body is a machine which winds its own springs merely by feeding itself.

The materialist conception of human nature had a wide following in the eighteenth century, especially among French thinkers who sought to base their study of man and society on a solid empirical foundation, patterned after the method of the physical sciences. Some of the most prominent philosophers—Diderot, Condillac, Helvétius, Holbach, and, with some reservations, even Rousseau—accepted a sensationist epistemology and the ethics and political theory that derived from its consistent application. They followed a general sequence of reasoning that went as follows: the mind is a blank tablet at birth and develops as a consequence of the action of pleasurable and painful sensations on it; man behaves so as to maximize the pleasurable sensations; man ought to act so as to maximize his pleasure; society ought to be organized so as to maximize the greatest pleasure for the greatest number; a science of society can be developed to reduce human suffering; and ultimately mankind must progress through the march of reason and the steady accumulation of knowledge.

The philosophical culmination of eighteenth-century sensationism appeared *in extremis* in the philosophy of the Marquis de Sade, who reasoned that if the pursuit of pleasure is the only reliable guide for human behavior, then the free indulgence of human sensuality must be man's highest goal. In a series of fantasies of sexual debauchery de Sade offered his philosophy of the goodness of unrestrained sexuality involving an exhaustive experimentation with bodily sensations of pleasure and pain.

In the early nineteenth century Schopenhauer's emphasis on the importance of will in human understanding of the world included an argument for the primacy of corporeality. Knowledge of the world is always given through the medium of the body. The will reveals to man the essence of his existence, and since acts of the will are executed by the body, man is dependent on his body both for knowledge of and action on the world. Schopenhauer further underlined the importance of human sexuality by interpreting the sexual impulse as "the ultimate goal of almost all human effort." [2] The unenlightened individual strives to serve the species, although he is deluded into thinking that he serves himself. And though Schopenhauer regarded this striving of the will as absurd, he nevertheless argued convincingly its central role in governing the affairs of mankind.

In the latter half of the nineteenth century, developments in medicine, physiology, psychology, and philosophy converged in a predominantly materialistic conception of human nature. The opposing view of the vitalists provided fuel for a debate that raged throughout the century. The materialistic view was defended by the mechanists, who argued that all processes operative in the human organism—both mental and physical—could ultimately be explained as the action of physio-chemical processes. One cluster of materialists centered in German universities and, under the influence of René Du Bois-Reymond, in 1842 formally pledged a "solemn oath to put into effect this truth: No other forces than the common physio-chemical ones are active within the organism. In those cases which cannot at the time be explained by these forces, one has either to find the specific way or form of their action by means of the physical-mathematical method or to assume new forces equal in dignity to the chemical-physical forces inherent in matter, reducible to the force of attraction and repulsion." [3]

In 1855 the German biologist Ludwig Büchner complained that theories of vital forces had "deeply injured the cause of science." He assailed these vitalistic explanations and substituted blatantly mechanical examples to illustrate the nature of the human body. "The animal stomach," he argued, "may correctly be described as a chemical retort." Digestion, respiration, and nervous activity have been explained as purely chemical processes; and "the heart is furnished with valves like a steam engine." The construction of the eye "rests on the same laws as does the construction of a *camera obscura*," and, he concluded, "every day it is becoming more obvious that *electricity*, a well-known natural force, plays a most important part in these organic phenomena." [4] The mid-century witnessed the ascendency of materialist theory, which, together with the positivist method of investigation, dominated the biological sciences until the end of the century.

In "On the Hypothesis that Animals Are Automata, and Its History" (1874) Thomas Huxley traced the historical origin of the ideas that the living body is a mechanism and that the physical processes of life can be explained like other physical phenomena. These views, he concluded, were the foundation for the current study of scientific physiology.[5] Huxley had predicted as early as 1863 that "even man's 'higher' abilities of intellect, feeling, and will could ultimately be explained as changes in the position of various body parts."[6] In 1868 Huxley articulated an extreme materialist position in an article titled "On the Physical Basis of Life," which argued that all living processes could ultimately be explained as the activity of "protoplasm." He further argued, playing on Descartes's machine metaphor, that "digestion, respiration, thought, imagination . . . in the machine naturally proceed from the mere arrangement of its organs, neither more nor less than do movements of a clock."[7] The struggle over the materialist view of life can be seen within the work of Huxley himself, who was torn between his youthful religious beliefs about the nature of man and the revelations of his own research. That work led him to champion Darwin's theory of evolution and explore the physio-chemical processes that appeared to explain so much of the human nature that had previously been attributed to the action of invisible substances and mysterious forces.

Another English biologist, G. J. Allman, addressed himself to the philosophical significance of Huxley's paper in 1879 in "Protoplasm and the Commonality of Life." Research in biology by that time had revealed that protoplasm was a life-sustaining substance common to both plants and animals. For a generation that was still having trouble digesting the theory of evolution, the argument that humans are fashioned out of the same stuff as plants was an unbearable assault on human dignity. Allman accepted the fact that there was a common tie between plants and animals in their vital dependence on protoplasm, but he insisted further, almost hysterically, that "between thought and the physical phenomena of matter . . . there is no conceivable analogy."[8]

While some were tortured by the disappearance of the

mysteries of life, others found it a source of ennoblement rather than degradation. The English essayist James Hinton greeted the prospect of explaining life in physio-chemical terms with glowing expectation. "Would it not be beautiful to see these forces stand before us in a new attitude and with more than doubled lustre?" [9] The materialist position places man in nature and removes the confusion and mystery of living organisms. The human mind and body is a complex of chemical and mechanical processes suspended in a state of dynamic tension which constitutes life. Nutrition and growth are produced by a continual struggle with decay and death. In the introduction to a reprinting of Hinton's *Life in Nature* in 1932, Havelock Ellis remarked how cold and ugly the universe had appeared to him as the older vision of man and the universe faded, to be replaced by the world of engineers and materialists. His reading of Hinton's book seemed to restore the glory and warmth of life. Instead of debasing life to the inorganic, Hinton elevated life to the organic, setting it in nature as part of a single unity of the organic and inorganic.

Another important contribution to the materialist conception of life was made by the French biologist Claude Bernard, who endeavored in 1865 "to prove that the science of vital phenomena must have the same foundations as the science of the phenomena of inorganic bodies." [10] Bernard rejected all vitalist explanations of organic activity. Life is a result of the action of physio-chemical processes, and the human body is regulated by an "inner physiological environment" (*milieu intérieur*). "A living organism is nothing but a wonderful machine endowed with the most marvelous properties and set going by means of the most complex and delicate mechanism." [11] The cultural impact of this book was partly the result of the particular emphasis that his phrasing "nothing but" gave his theory. The mechanists of this period took the offensive and assaulted vitalists and idealists. They insisted that mankind must learn to do without the theoretical ornamentation that obstructed the empirical study of man. Traditional views of the nature of human existence had denied its corporeal underpinnings and hence had not encouraged the study of the complex chemical processes that produced and sustained life. And so Bernard sharpened his argument by insisting that a living organism is "nothing but" a wonderful machine.

A study on the nature of love that appeared in Boston in 1869 by a Rosicrucian, the Count de St. Léon, documents the genre of popularizing and overextending the modest findings of the physiologists. The era of positivist science is dawning: "People now begin to understand of what their bodies are composed." Mental and physical existence "are but so many chemical changes." "A few years ago it was discovered that all living creatures were made up of . . . minute bodies called cells." Then Huxley argued that protoplasm was the basis of life. Physiologists reduced living beings to organizations of contracting and pulsating matter, and now we know that the entire mental and moral sphere of human existence is governed by bodily processes. "If you want to feel happy," he advised, "look after your digestive and circulatory systems." Women are particularly governed by their bodies, what with their cramped waists, contracted lungs, and fevered stomachs. "Pelvic inflammation is the national disease," and the environment is filled with disease-causing "spores, parasites, and animalculae." The significance of the book's title, *Love and Its Hidden History*, is revealed as he explains the importance of human sexuality. Twenty-five years before Freud, the Count de St. Léon argued that "all nervous diseases spring from disarrangements of the sexual system." He added that "love is not only liable to, but often is, the subject of disease, and from the diseases thus originated spring nine-tenths of all human ailments." [12] He concluded with a detailed account of nutrition, metabolism, and other bodily processes—all part of the hidden history of human love.

While the biologists and physiologists argued the physiochemical origins of life, physicians and philosophers explored the interdependence of the mind and the body and simultaneously pursued modern psychosomatic medicine and the philosophical critique of Cartesian mind-body dualism. The traditional division of human existence into two separate realms of the mental and the physical was substantially reconsidered as the mutual interaction of mind and body was closely scrutinized in the latter decades of the century. I shall consider the attack on Cartesian dualism again in Chapter 18 on the philosophy of the body.

The English physician William Hooker attempted to steer a middle course between dualism and materialism as he viewed them in 1850. On the one hand, he complained that most physicians tended to overlook the influence of the mind on diseases of the body. The dualist notion that the mind dwells in the body "does not express the extent nor the intimacy of the connection," because the mind is bound to the body in its every nerve and fiber. On the other hand, the materialists err in thinking that "the brain produces thought, pretty much as the liver makes bile or the stomach gastric juice." It is nevertheless essential for physicians to recognize that "all the manifestations of mind are not only connected with, but are dependent upon, a material organization." [13] The wise physician must be aware of the mental state of his patient—especially his fears—which may influence the success of the treatment.

The prominent English physician Henry Maudsley pioneered psychosomatic medicine in a number of influential works beginning in the 1860s.[14] In *Body and Mind* he explained physical and mental interaction and elaborated upon the theory of organic memory according to which memories involve the mental registration of bodily experiences. Hence, efforts of the mind may fail to revive memories that a fever, a blow on the head, or a smell will recall. Sexual activity involves a most intimate interaction of mind and body such that "when an individual is sexually mutilated at an early age, he is emasculated morally as well as physically." Eunuchs are cowardly and deceitful, joyful anticipation affects breathing, the consumptive is sanguine, and fear strikes the heart. Emotions directly affect digestion and even nutrition. "Neither in health nor in disease is the mind imprisoned in one corner of the body; and, when a person is a lunatic, he is . . . lunatic to his fingers' ends." Maudsley concluded with a comment intended to take up objections against his materialist leanings. It offers a glimpse at the kinds of fears generated by the extension of human corporeality into spheres of existence traditionally regarded as exclusively governed by the mind. He wrote: "I have no wish whatever to exalt unduly the body; I have, if possible, still less desire to degrade the mind; but I do protest . . . against the unjust and most unscientific practice of declaring the body vile and despicable, of looking down upon the highest and most wonderful contrivance of creative skill as something of which man dare venture to feel ashamed." [15]

This appeal from within the medical profession echoed many of the developments in the arts and literature of the period. A number of philosophers attempted to explain aesthetic response in purely materialistic terms. A pathbreaking contribution to physiological aesthetics was made by Hermann Helmholtz, whose *Sensations of Tone* (1862) presented a theory of harmony based on the physiology of the human hearing apparatus. He demonstrated that the quality of a tone is determined by the order, number and intensity of the harmonics of a sequence of sounds which resonate in such a way as to please the human ear. Another strongly materialistic aesthetic was announced by another German experimental psychologist, Gustav Fechner, whose "Experimental Aesthetics" of 1871 reduced the aesthetic response to pleasure and pain—with beautiful objects arousing a preponderance of pleasure.[16] He argued that aesthetically pleasing visual or auditory sensations share three qualities: they are clear, they are consistent, and they unify manifold material. Grant Allen argued along a similar line that there is a "purely physical origin of the sense of beauty." He speculated that "pain is the subjective concomitant of destructive action or insufficient nutrition in any sentient tissue" while "pleasure is the subjective concomitant of the normal amount of function in any tissue." [17] Judgments about beauty are physiological responses which either serve in the interest of building up an organism or tearing it down. Throughout history those objects that have contributed to the evolution of the species have come to be regarded as beautiful. They confirm higher development and thereby generate pleasure. Allen's book is of historical interest because it demonstrates how uncompromisingly one aesthetician at that time sought to reject the idealist aesthetic for a purely physiological one.

The culmination of the movement to found a materialistic aesthetic came from George Santayana in 1896. He began by inquiring what elements of human nature make man sensitive to beauty; his inquiries led him to the conclusion that the pleasures of enjoying beauty differ from purely physical pleasure because the aesthetic involves "impersonality" of the enjoyment. Hence the painter does not look at a fountain as does a thirsty man, or at a beautiful woman as does a satyr. Nevertheless pleasure does play a role; it "lends to the visible world that mysterious and subtle charm which we call beauty." The condition of the body directly influences our appreciation of beauty, and without health no pleasure can be pure. For example, the pleasure of breathing is essential to our most transcendental ideals. "It is not merely a metaphor that makes us couple airiness with exquisiteness and breathlessness with awe; it is the actual recurrence of a sensation in the throat and lungs that gives those impressions an immediate power." Santayana concluded with a discussion of the way the sexual instinct influences both artistic creation and artistic response. The theory that art is sublimated sexuality has a long history, particularly in the nineteenth century, and neither Freud's nor Santayana's version of it was entirely original.[18]

The rise of Christian Science forms another part of the growing interest in the interaction of body and mind. In 1866 Mary Baker Eddy "discovered" Christian Science. It derived from the fundamental insight that proper faith can maintain physical as well as mental health. She concluded that the mind governs the body and that false belief generates disease. "Human mind produces what is termed organic disease as certainly as it produces hysteria," and her method of healing would involve, therefore, the restoration of a proper mental state—true Christian faith. At a time when physicians were still bleeding patients, when obstetricians continued to spread puerperal fever from woman to woman in maternity wards, and when quacks persisted in showering toxic medicinal potions on the public, Mary Baker Eddy's insistence that most physicians created more disease than they cured had a strong appeal. Though her emphasis was on the influence of mind over body, the basic message of her medical-theological system contributed to the general cultural awareness of the interaction of body and mind.

The precise nature of the interaction of the mind and the body was pursued more exactingly by a group of psychologists in the nineteenth century who believed that the brain was the seat of the mind and that its various "functions" or "faculties" were localized in specific anatomical regions of it. This study, phrenology, was pioneered in the late eighteenth century by the French anatomist Franz Joseph Gall, who believed he had discovered a connection between mental abilities and the shape of the head. Various "bumps" on the head indicated enlargement of the immediate underlying part of the brain, and those enlargements, Gall believed, further indicated that the particular faculty localized in that region was highly developed. His student, the Viennese physician Johann Spurzheim, elaborated on the theory and developed the doctrine of physiognomy, which purported to interpret character and mental ability from a study of the face and skull. The system developed by Gall and Spurzheim included thirty-seven faculties of the mind, and phrenological diagrams charted the precise region where these faculties resided.

The idea that character is reflected in the outer configurations of the head appealed to many who sought some clue to unravel the mysteries of human personality. The appearance of precise anatomical precision in phrenology camouflaged the arbitrariness of the theory. Consider one explanation of the location of "Secretiveness" by the famous American phrenologist F. N. Fowler in 1881. "Secretiveness is located ¾ of an inch above the middle of the top of the ears. . . . When the head widens rapidly from the junction of the ears as you rise upward, Secretiveness is larger than Destructiveness; but when the head becomes narrower as you rise, it is smaller than Destructiveness." [19] To the believer, the empirical evidence in favor of the theory was sufficient. Criminals had low foreheads, and men of great intellect and moral strength had high ones.

Darwin's study of character, *The Expression of Emotions in Man and Animals* (1872), seemed to offer the highest scholarly credentials to phrenology, because in it he argued that the emotions are expressed in bodily movements. Shrugging the shoulders was a sign of helplessness, raising the arms with the hands open was a sign of wonder. Darwin was intent upon showing two facts: that some expressions are universal among the different races and are therefore innate and a derivate of ancestral prehistory, and second, that there was a similarity between the emotions of animals and man, which was further evidence for his theory of the evolution of man from lower animal forms. He did not endorse the wild speculations of the phrenologists, even though his book was frequently cited as scholarly corroboration by them.

The reading of personality in the configuration of facial features gained wide interest as a corollary cultural development to phrenology. Physiognomists presented a number of social and ethnic prejudices under the pseudo-scientific gloss of phrenological rhetoric. The eye was the most reliable index to personality, but some physiognomists ventured to interpret the characterological significance on the nose. Alfred T. Story's essay of 1881, "A Chapter on Noses," explained that noses turned downward indicate a sarcastic, ill-tempered, and hypochondriacal temperament. There are five kinds of noses. The Roman nose reflects executive and aggressive qualities, while the Greek nose indicates an artistic temperament. A "snub" nose reveals an underdeveloped character, and the upturned or "celestial" nose shows an inquiring and receptive mind. The "Jewish" nose indicates a keen, apprehensive, and suspicious type; it "betokens a disposition to make schemes and study men." [20] Another theorist of noses offered a similar explanation of the character indicated by the "Jewish" or "hawk" nose: "commercial spirit, love of making money and a desire to accumulate." [21] These speculations on noses anticipated a later ill-fated chapter in the history of phrenology in the twentieth century—the resurgence of interest in phrenology in Germany in the 1930s, when Nazi theoreticians scanned the past century of European culture for evidence to support their racial theories. I shall take up the revival of phrenology in Chapter 17 on body politics in Germany.

The introduction of anesthesia for the alleviation of pain in surgery generated further interest in the relation of the mind and body and triggered a fierce debate at mid-century over the moral and religious justification for relieving human suffering. Pain produces the most immediate evidence for the sensitivity of human life. Its function as a warning that the body is in danger is a reminder that our existence is constantly threatened by death. Even more vividly than pleasure, pain brings man squarely up against his corporeal nature. That intimate connection between physical suffering and life became the subject of a controversy that began with the introduction of anesthesia to deaden the pain of surgery, particularly of childbirth.

In 1795 Sir Humphrey Davy discovered that the inhalation of nitrous oxide produced giddiness, and in 1818 Michael Faraday found that ether produced similar effects. Throughout the first half of the nineteenth century both gases were used for entertainment at "laughing-gas parties" or "ether frolics," until in 1842 a dentist from Georgia, Crawford Long, experimented with the use of ether to induce anesthesia for a dental operation. A Massachusetts dentist, William Morton, first published on the use of ether, and in 1847 a Scottish physician, Sir James Simpson, delivered a child to a woman anesthetized with chloroform. The Calvinist clergy of Edinburgh soon raised a cry of protest against Simpson by arguing that God had intended woman to endure the pain of childbirth. The Bible says that "in sorrow thou shalt bring forth children," and "sorrow" was interpreted to mean "pain." [22] Some argued that the pain was God's punishment for the sin of Eve or that it was a reminder of the punishment for the pleasures of the flesh. One Boston physician insisted that the suffering of a woman in labor "is one of the strongest elements in the love she bears for her offspring." [23] The alleviation of the agony of childbirth seemed to be contrary to the will of God, who had ordained that women must suffer to bear a child. Protest against the use of anesthesia for women in childbirth was substantially subdued when Queen Victoria took it to deliver her seventh child, Leopold, in 1853.

Although I have little solid evidence to support the following speculation, I believe it is correct to say that the possibility of alleviating physical pain profoundly altered the going view of the "value" of pain and lowered the estimation of asceticism. Christian asceticism has often maintained that suffering ennobles life. One Christian thinker wrote in 1877 that pain "clarified and purified life" and educated the soul in preparing it to face death.[24] In an essay on pain, Jules Rochard concluded that following the introduction of anesthesia to reduce pain, Europeans had grown to fear pain more than death and had become less able to endure suffering. "The suppression of pain has exaggerated the sensibility that one observes today particularly among the upper classes." [25] He added that the experience of pleasure is also more delicate than it was for his ancestors. With some regret he concluded, however, that as a consequence of the refinement of life and the heightened sensitivity to pleasure and pain, the ability to endure pain and to brave danger had been lost. It is also possible that there may be some connection between the elimination of public executions (the last public hanging in England was in 1868) and a greater sensitivity to human suffering.

I would speculate that this heightened sensitivity to pain, like the heightened sensitivity to smell, was part of a general cultural reorientation toward bodily existence. The possibility of reducing pain and of controlling body odors and environmental stench allowed a more refined human physical response to these

MATERIALISM AND THE MIND-BODY PROBLEM   79

kinds of stimuli. Cultivation of physical sensibility is only possible when the extremes of stimulation are subdued to allow refined response. Gourmets do not thrive during famine.

I would like to conclude this chapter with some speculations about the possible influence that mechanical technology may have had on body image. The late nineteenth century was an age in which machines came to replace human labor, with a consequent transformation of both the productive process and the men working in it. The machines that replaced laboring human bodies began to reshape human nature—they were extensions of the body and magnified human labor potential. But there also occurred an opposite action of the machine which constricted the human body. As men competed with machines they began to imitate them. The body began to lose its spontaneity as it fell into line with the pace of factory production. Lewis Mumford has interpreted the effect of science and technology as leading to a "sterilization of the self" and an elimination of the human pleasure in man's own image. He compared the impact of technology with the monastic ideal—"elimination of the body, of human sensuous involvement" in life. In contrast "older cultures tended to emphasize respect for the body and to dwell on its beauties and delights." [26]

While Mumford focused on the constricting and sterilizing impact of machines on human body image, Marshall McLuhan has emphasized their expanding potential. During the mechanical age, McLuhan argued, the human body was extended in space and its powers were magnified and exalted. These two interpretations of the impact of machine technology embrace dialectically related contrasting views of the changing conception of human nature witnessed under the rapid transformation of human corporeal existence in the last century. Machines have both constricted and extended the human body. The effects may have appeared either way to different critics, but the potential for both influences was there. The general cultural impact of these opposing influences on body image was a growing awareness of the nature of the body and its responsiveness to technological change.

Chapter 8

The Body Electric

The challenge to Victorian sexual morality, we have seen, began around 1850, and thereafter the demand for a fuller indulgence of sexuality increased. Walt Whitman's *Leaves of Grass* (1855) offered a vision of the sexually unrestricted natural man. "Song of Myself" eulogized his own body and those of his many lovers—male and female. "Welcome is every organ and attribute of me, and of any man / hearty and clean, / Not an inch nor a particle of an inch is vile, and none shall be less familar than the rest." His response to nature was charged with sexuality, and he talks of running his fingers through the blades of grass with the same intensity as stroking his lovers' hair.

Whitman did not come to this free sexuality with ease, and his poetry records the struggle to cast aside "all the standards hitherto publish'd." He worked to release the sensuality that had been bottled up for so long. "From pent-up aching rivers" he learned to give way to his impulses and wrote, "From my own voice resonant, singing the phallus, singing the song of procreation."

The longest single poem on the body, "I Sing the Body Electric," is a series of images of naked, athletic, beautiful human bodies offering themselves to one another without restraint in nature—in the grass, behind a rock, by a lake. "If anything is sacred," he wrote, "the human body is sacred." He concluded with a detailed description of the various parts of the body from the head to the heel to emphasize his responsiveness to every square inch of skin, the movement of every muscle and nerve beneath, even the odors and exhalations. Midway through this strange poetic-anatomical digression he wrote: "Strong shoulders, manly beard, scapula, hind-shoulders, and the ample side-round of the chest, / Upper armpit, armpit, elbow-socket, lower-arm, arm-sinews, arm-bones"—a catalog of the wondrous shapes and movements that physically encase the human soul. He concluded: "O I say these are not the parts and poems of the body only, but of the soul, / O I say now these are the soul!"

"A Woman Waits for Me" explored the comprehensiveness of human sexuality:

> Sex contains all, bodies, souls,
> Meanings, proofs, purities, delicacies, results, promulgations,
> Songs, commands, health, pride, the maternal mystery, the seminal milk,
> All hopes, benefactions, bestowals, all the passions, loves, beauties, delights of the earth,
> All the governments, judges, gods, follow'd persons of the earth,
> These are contain'd in sex as parts of itself and justifications of itself.

Whitman's popularity among homosexuals later in the century can be explained by the numerous poems which described his homosexual loves. Although he was apparently bisexual, his references to homosexuality reveal that some of his most intense passions and his most bitter disappointments involved his relations with other men. Western culture did not get as explicit a public confession of homosexuality from a prominent intellectual again until Gide confessed in 1926 after years of indecision and concealment.

As Whitman had attempted to clear the congestion that choked human sexuality in America, his European contemporaries were beginning a similar revolution in the conventions that limited explicitly sexual imagery of poetry and literature. Baudelaire challenged many romantic conventions about human sexuality in *The Flowers of Evil*, and in England the Pre-Raphaelites, led by Dante Gabriel Rossetti, were creating a most sensuous poetry. One English critic made a name for himself with a passionate denunciation of the "fleshly poets" of the age. Robert Buchanan's *The Fleshly School of Poetry* (1872) catalogued the fears which explicit human sexuality in art generated in a society committed to sexual repression. He argued that "sensualism" threatened all societies, but, beginning with de Sade and revived by Baudelaire, it was becoming particularly dangerous. The streets were full of shops that flaunted pornographic photographs in their windows. "It has penetrated into the very sweetshops; and there . . . may be seen this year models of the female Leg, the whole definite and elegant article as far as the thigh, with a fringe of paper cut in imitation of the female drawers and embroidered in the female fashion! . . . The Leg, as a disease . . . becomes a spectre, a portent, a mania . . . everywhere—the Can-Can, in shop-windows." Baudelaire and Rossetti, Buchanan claimed, were particularly responsible for this "Scrofulous School of Literature" which took the hideous for its subject. Baudelaire's only merit "was in his nasal appreciation of foul odours." Rossetti was guilty of putting on record "the most secret mysteries of sexual connection." His females "bite, scratch, scream, bubble, munch, sweat, writhe, twist, wriggle, foam, and in a general way slaver over their lovers." Buchanan protested that he had never encountered women who conducted themselves in such a manner, and he was particularly outraged by references to oral sex. Buchanan repeated toward the end of the book (and one can imagine his shaking his head as he did so) that he had never met such women as Rossetti described, and, in a final burst of abuse, accused Rossetti of producing "effeminate poetry."[1] These complaints were echoed throughout the century as writers continued to describe the body in increasingly vivid detail. The focus of this type of criticism in the latter decades of the century was the work of Emile Zola.

In a series of novels about the Rougon-Macquart family Zola reconstructed many of the fantasies and emotions that dominated European sexuality at the time. Zola's thoughts about human sexuality revealed a central confusion and a central conflict, fluctuating according to his changing theoretical interest and emotional state. The confusion was about the nature of hereditary transmission of character traits and the ways in which people influence each other biologically. Zola was much impressed by Claude Bernard's theory of the way various fluids of the "interior milieu" of the body regulate life. Zola viewed himself as experimenting with characters upon whom the influence of heredity and environment worked according to discoverable laws. He intended to create a family, fully describe its hereditary endowment, and then observe with the objectivity of a medical experimenter the interaction of the laws of heredity and those of society upon the family. The "experimental novelist," he wrote, "must work with characters, emotions, human and social facts, as the chemist works with matter, as the physiologist works with living bodies." [2] In his early novels he was preoccupied with the ways in which people influence one another by bodily fluids. In *Thérèse Raquin* he introduced his theory of temperaments, which held that a man and wife influence each other's temperament over the years by the transmission of subtle substances that they exchange by kissing, breathing, and having sexual intercourse. "These modifications," he wrote, "which have their origin in the flesh, are speedily communicated to the brain and affect the entire individual." [3] This strongly materialistic view of the physio-chemical determinants of temperament was coupled in his mind with an equally materialistic theory of impregnation and hereditary transmission. In another early novel, *Madeleine Férat*, Zola argued that a woman who has cohabitated with a man remains tied to him by physiological bonds, and every subsequent child she may bear to another man will in some ways resemble her first lover. Madeleine, he explained, was ignorant of the "fatalities of the flesh" that may bind a woman to her first lover. Thus parents and children, husbands and wives, are bound to one another forever as a consequence of the exchange of substances transmitted by sexual intercourse.

In addition to this confusing conception of hereditary transmission and of the capacity of sexual substances to influence the temperament of another individual, Zola was torn by a conflict about the morality of human sexuality. His attitude fluctuated between an earlier prudishness and a subsequent reverence for the life-affirming capacity of sexual pleasure. In *La Confession de Claude* he revealed his fantasy that his wife should come to him unspoiled from the hand of God. "I should like her white, pure, not yet living, and I should awaken her!" [4] This attitude led him to experience the most intense passions, which he described in a number of novels. *La Bête humain* opens with a scene of fanatical jealousy, the agony of Count Muffat in *Nana* is described at length, the driving force in *Une Page d'amour* is a daughter's suspicion of her mother's lover, and the envy the rich experience over the sexual libertinage they believe the lower classes enjoy adds to the class conflict explored in *Germinal*. The anguish of the manager Hennebeau no doubt rang true to many of Zola's readers who themselves felt trapped within the restrictions of their sexual morality. Hennebeau reflected resentfully on the rebellion of the miners: "How gladly he would have given them his fat salary to have tough skin like them and copulate easily and without regret. If only he could have sat them at his table and stuffed them with his pheasant, while he went off to fornicate behind the hedges, laying the girls and joking about those who had laid them before him!" [5] The only physical contact that does take place between the workers and owners of the mines in *Germinal* occurs when the worker Bonnemort strangles the daughter of one of the mine owners.

Zola explored the conflict over sexual morality in an early novel in the Rougon-Macquart series, *La Faute de l'Abbé Mouret*. To become a priest Mouret had to conquer the stirrings of sensuality that continually threatened to distract him from his religious calling. He developed a "horror of physical sensation" requisite to maintain his strict celibacy, but a chance meeting with the sixteen-year-old Albine revived his long-suppressed sexuality. When he finally succumbed to Albine's temptings, the entirety of nature echoed its pleasure. The garden where he and Albine make love comes to life in Zola's lush prose to "share the couple's orgasm in one last cry of passion." The trunks of trees bend as if to nod approval, and "grasses released drunken sobs." Mouret discovers that sex has restored his manhood and his health, and for a brief moment he and Albine enjoy the full pleasure of sexual love. But they are soon driven apart by guilt. Albine worries that he will resent his dependency on her, then begins to fear that they were observed. She urges that they hide, and when she realizes that she is naked before him, feels ashamed. Mouret begins to share her shame, and as they emerge from the garden and look upon the

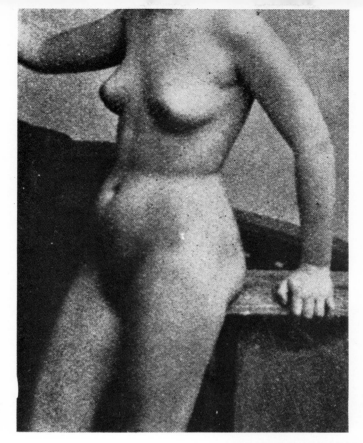

1. Permanent bodily disfiguration from tight lacing. Reproduced from DIE KULTUR DES WEIBLICHEN KÖRPERS ALS GRUNDLAGE DER FRAUENKLEIDUNG by Paul Schultze-Naumberg, Jena, 1922.

2. Woman suffering from spinal curvature created by prolonged wearing of the "health" corset introduced around 1900. Reproduced from DIE KULTUR DES WEIBLICHEN KÖRPERS ALS GRUNDLAGE DER FRAUENKLEIDUNG by Paul Schultze-Naumberg, Jena, 1922.

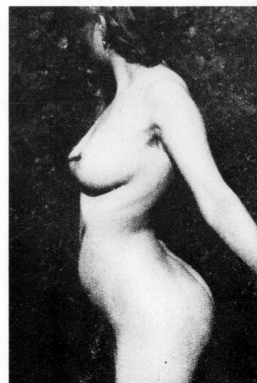

3. François Gérard, *Psyché recevant le premier baiser de l'Amour*, Paris, Musée du Louvre, courtesy Photographie Giraudon.

4. Jean Ingres, *La Source*, Paris, Musée du Louvre, courtesy Caisse Nationale des Monuments Historiques.

5. Jean Ingres, *Le Bain Turc*, Paris, Musée du Louvre, courtesy Photographie Giraudon.

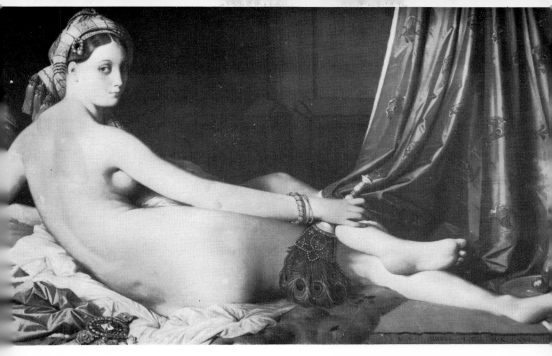

6. Jean Ingres, *La Grande Odalisque*, Paris, Musée du Louvre, courtesy Caisse Nationale des Monuments Historiques.

7. Édouard Manet, *Olympia*, Paris, Musée du Louvre, courtesy Agraci-Paris.

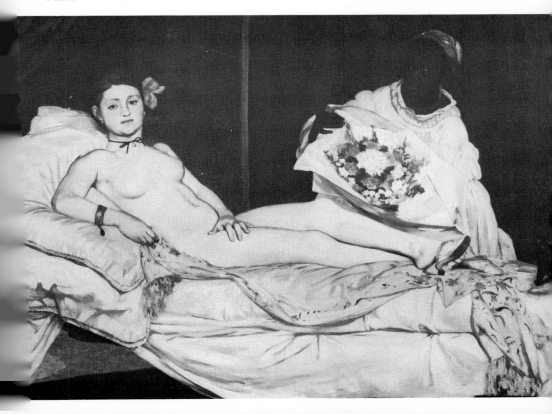

8. Théodore Géricault, *Le Radeau de la Méduse*, Paris, Musée du Louvre, courtesy Cliché des Musées Nationaux.

Théodore Géricault, *Study of a Nude Man*, New York, Metropolitan Museum of Art, Rogers Fund, 1952.

10. Édouard Manet, *Le Dejeuner sur l'herbe*, Paris, Musée du Louvre, courtesy Agraci-Paris.

village whose morality they have just offended, both are thrown into despair. The sound of church bells intensifies Mouret's guilt, and by the time Brother Archangel emerges from the bushes castigating them both for their wickedness Mouret is ready to renounce Albine and return to the priesthood. Giving himself to a fantasy of running away with Albine, he is suddenly stopped short at the thought of having children. "He could not get used to this flesh of his flesh, which seemed all sweaty with his man's impurity." The conflict ends unresolved as he remains in the priesthood and she commits suicide.

Zola's most thorough study of the power of sexuality was his novel about a Parisian prostitute, *Nana* (1880). The opening chapter established the awesomeness of this power as Zola described the effect on the audience produced by Nana's appearance on the stage. "She was naked with a calm audacity, certain of the supreme power of her flesh. . . . And when Nana raised her arms, one could see the golden hairs in her armpits reflected in the glare of the footlights. No one applauded and the laughing stopped. The men leaned forward with serious faces. Their nostrils contracted, their mouths became inflamed and parched. . . . Suddenly, out of the virtuous child the woman arose, bearing the insanity of her sexuality, opening the unknown world of desire. Nana was smiling still, but it was the bitter smile of a man-eater." As the account of Nana's reputation proceeds, Zola reveals his own apprehension about the enormous force of sexuality. "Her slightest movements blew the flame of desire; she ruled men's flesh with the flick of a finger." [6] That sexual power enables her to ruin a host of admirers from the highest to the lowest ranks of Parisian society. And as she does so, it becomes clear that she is also a victim of the corrupting powers of her own sexuality. She was a "leaven of destruction" who poisoned everyone who drew near. One of her lovers writes a story which describes Nana's life. "She was descended from four or five generations of drunkards and tainted in her blood by a cumulative inheritance of misery and drink, which in her case was transformed into a nervous exaggeration of the sexual instinct. She shot up to womanhood in the slums and on the pavements of Paris. Tall, beautiful, with superb flesh like a dunghill plant, she avenged the beggars and outcasts of whom she was the product. With her the decay that is allowed to ferment in the people has returned and rots the aristocracy. She became a force of nature, a leaven of destruction, and unwittingly corrupted and disorganized Paris between her snowy thighs, churning it as housewives churn milk each month." [7] As another of her admirers reads this article aloud, Nana parades before a mirror, lost in "ecstatic self-contemplation." The beauty of her body, its power to control and destroy, has amazed even her. She views her own body almost as if it does not belong to her, posing and turning before a mirror.

Surveying the field of her ruined lovers, Zola concludes with the judgment that "it was well done—it was just." Zola views her "labor of ruin and death" as a fitting revenge for the injustices done to the members of the lower classes—the beggars and wastrels from whose midst Nana arose. She, however, remained unconscious of the complete victory of her sex over Parisian society.

The destructive power of her sexuality finally completes its labor of ruin upon Nana herself. During a holiday Nana contracts smallpox, which slowly disfigures her face and strips her of the power that had manipulated the fate of so many. "Venus was rotting. It seemed as though the virus she had absorbed from the gutters, . . . the leaven with which she had poisoned a whole people, had remounted to her face and turned it to decay." [8] Zola leaves no doubt that he regards sexuality as a wild and potentially destructive force, but his vivid descriptions of Nana's body, his detailed accounts of the life of this courtesan, his scrutiny of the fate of men struck down by lust reveal his fascination with sex. His descriptions of her body excited a generation of readers at a time when literary conventions excluded scenes suggesting sensuality, and when the final lines of love stories concluded with a first embrace. Zola gave the fashionable literary world a graphic account of what happened *after* "she sank into his arms," and in doing so generated a storm of critical protest.[9]

Though Zola's view of sexuality in *Nana* appears to be largely negative, it nevertheless reveals traces of that other attitude he assumed when writing about the beauty of the sexual embrace. Though Zola takes Nana through a series of disastrous episodes caused by her seductive powers, he also appears to admire her and be strongly attracted to the incarnation of female sexuality he has created. In an age of repression she symbolized all of the rough-and-tumble sexuality that was denied to the "respectable" classes. But when they got going, Zola pointed out, those classes could display a wide range of sexual aberrations, including even a taste for masochism and bestiality. "From one end of the social ladder to the other, everybody was on the prowl." [10] The supreme triumph of her class is achieved by having the Count Muffat stamp on his coat, spit on his medals, and then crawl about on all fours and fetch a handkerchief in his mouth like a dog.

Zola was interested in the full range of human sensuality and corporeality beyond the realm of the exclusively sexual. We have examined his preoccupation with smells. His novels are filled with vivid accounts of a variety of human bodily functions that no doubt shocked his readers as much as the sexuality. The earthy sensuousness of the peasants in *La Terre* spills over into rapes, beatings, and one of the most savage murder scenes in nineteenth-century literature. Old Fouan, who has divided his land between his children, becomes a burden to each, and eventually his ward, Lisa, decides to kill him. She thrusts a pillow over him during his sleep, but only breaks his nose. To conceal her crime she, together with her accomplice Buteau, decides to burn him alive and sets fire to his beard. This revives him for a moment, and, with his face blackened and his beard on fire, he gives them a dying look of hatred and agony. They finish him off by setting fire to his mattress. Probably the most startling scene in all of Zola's novels is the castration of the owner of the company store in *Germinal*. The local mine workers, crazed by frustration and hunger during a hopeless strike, go on a rampage that eventually takes them to the company store. The hated owner, Maigrat, who occasionally took payment for groceries in sexual pleasures with the miners' daughters, falls from the roof of his store and is killed. One enraged woman, "with her withered hands, separated his naked thighs and grabbed his dead virility. She took hold of the lot and pulled with an effort that strained her skinny back and made the bones crack in her long arms. The soft skin resisted and she had to try again, but finally she succeeded in tearing away the lump of hairy and bleeding flesh, which she waved aloft with a triumphant laugh. . . . Then Ma Brûlé stuck the whole thing on the end of her stick and carried it down the street like a standard, followed by a rout of shrieking women." [11] At times Zola kept true to his original wish to work like an anatomist of living human bodies. In response to the complaint that the bodies of men and women were not limited to face and hands, Zola exposed the rest for public examination. People do have stomachs and sex organs, and underneath the skin there are nerves and muscles, all of which determine behavior. Against the one-sidedly cerebral view of human nature of the age, Zola pitted the full rage of corporeal existence.

Central to Nietzsche's aesthetics is his appeal for the cultivation of human sensuous existence without loss of the creative powers that require discipline and self-denial. A major concern of his philosophy is, therefore, the problem of how asceticism fits into the general structure of human consciousness. In the conclusion to his attempt in *The Genealogy of Morals* to retrace its rise, Nietzsche explained the significance of the ascetic ideal: it was a "hatred of humanity, of animality, of inert matter; this loathing of the senses, of reason even; this fear of beauty and happiness; this longing to escape from illusion, change, becoming, death, and from longing itself. It signifies, let us have the courage to face it, a will to nothingness, a revulsion from life, and rebellion against the principal conditions of living." [12] This argument was an attempt to point out a stage in the development of modern morals which involved an embracing of self-denial. His personal and philosophical struggles were devoted to trying to inquire how one could transcend that asceticism and affirm life once again.

In *The Birth of Tragedy* (1871) Nietzsche began his exploration of the problem of asceticism as an obstacle to artistic excellence. The complete work of art must achieve a reconciliation between two components of human existence—the Apollonian and the Dionysian. Nietzsche investigated the nature of these two spirits as they exist separately and treated their union first in Greek tragedy and then, as he discovered them once again reunited, in the operas of Richard Wagner.

The Dionysian spirit is akin to intoxication, and under its influence one experiences "the desire to sink back into the original oneness of nature." Music is the art form that most directly inspires it. The true Dionysian gratifies instinctual urges immediately and fully. The drunken frenzy of the Dionysian spirit, however, cannot itself produce great art. The reasoned and studied control of the Apollonian spirit is needed to give form and discipline to elemental passion. Wagner accomplished this control in his operas, in which the Apollonian spirit "parades the images of life before us and incites us to seize their ideational essence. . . . The Apollonian spirit wrests man from his Dionysiac self-destruction." [13] Already Nietzsche had tempered his criticism of asceticism with the idea that some sort of discipline is requisite to the production of great art. A central conflict of his thought remained the antithesis between the full gratification of man's instinctuality, the ability to enjoy the sensuousness of existence, and the need to control that instinctuality in order to give meaning to existence. Nietzsche was aware that instinct alone produces neither the good, the true, nor the beautiful. Both extreme asceticism and unchecked hedonism shut off essential elements of life and preclude the possibility of creative excellence. The dilemma confronting the artist or the philosopher is that while the union of reason and passion is essential to greatness, they summon forces which are in conflict with one another.

Nietzsche's critique of academic philosophy hinged partly on the ascetic implications of its leading doctrines. In *Beyond Good and Evil* he criticized the asceticism inherent in the idealist tradition, which repudiated the sensuous basis of human knowledge. The idealist asks, "How could anything originate out of its opposite?" How can one derive "truth out of error, or the will to truth out of the will to deception . . . or the pure and sunlike gaze of the sage out of lust? Such origins are impossible. . . . The things of highest value must have another *peculiar* origin—they cannot be derived from this transitory, seductive, deceptive, paltry world, from this turmoil of delusion and lust. Rather from the lap of Being, the intransitory, the hidden god, the 'thing-in-itself'—there must be their basis, and nowhere else." [14] This perpetual seeking for the essence of things beyond the world of appearances undermined faith in the senses, which Nietzsche insisted were the only source of our knowledge. The metaphysicians distrust their senses, "they rank the credibility of their own bodies about as low as the credibility of the visual evidence that 'the earth stands still,' and thus . . . let their most secure possession go (for in what does one at present believe more firmly than in one's body?)." [15]

Nietzsche attacked the aesthetics of Kant for similar reasons: ". . . the whole morality of self-denial must be questioned mercilessly and taken to court—no less than the aesthetics of 'contemplation devoid of all interest,' which is used today as a seductive guise for the emasculation of all art, to give it a good conscience." [16] Philosophers have generally misunderstood the body, and their efforts to provide a foundation for their thought independently of it have led to error. In *The Genealogy* he wrote: "The more emotions we allow to speak in a given matter . . . the more complete . . . our conception of it, the greater our objectivity." The full employment of the senses and the emotions is essential to our fullest understanding and enjoyment of the world.

Nietzsche's philosophy was poised between sharply contrasting views of several crucial philosophical issues. Before continuing with his attack on asceticism, it might be well to survey briefly some of his positive evaluations of it.

In the opening parable of *Thus Spake Zarathustra* Nietzsche explained the three metamorphoses that the spirit must go through. It must first be like a camel and learn to bear a load. In the second metamorphosis "the spirit becomes a lion who would conquer his freedom and be master in his own desert." This enables the spirit eventually to say "no" to the old values and to begin to exercise its freedom to create new ones. Finally the spirit must become like a child and learn to recapture its innocence and affirm a new world, a new set of values. The spirit must thus learn to utter "a sacred 'Yes' " and conquer its own world. A person must learn to bear a load, to discipline himself, and then, in the second stage, to reject the old values in order to be able to reaffirm the world of his own conception. Thus asceticism plays an essential role in the development of the mind as a negation which thereby constitutes a first step toward self-recognition. This constructive role of the ascetic mode of existence is elaborated in *Beyond Good and Evil*, where Nietzsche wrote: "What is essential . . . seems to be . . . that there should be obedience over a long period of time and in a single direction: given that, something always develops . . . for whose sake it is worth while to live on earth; for example, virtue, art, music, dance, reason, spirituality." He then recounted the many accomplishments that have been produced by this discipline, even though "in the process an irreplaceable amount of strength and spirit had to be crushed, stifled, and ruined." [17] But the central task of becoming an overman is a process of "self-overcoming" which involves learning how to endure pain. In some passages Nietzsche spoke of asceticism with genuine admiration: "The discipline of suffering, of *great* suffering—do you not know that only *this* discipline has created all enhancements of man so far?" [18] The special tension of the unhappy soul cultivates its strength, its ability to endure, to understand, and to create. In *The Genealogy* Nietzsche again considered the positive function of asceticism, but also warned of its dangers. The ascetic ideal saved man from suicidal nihilism, but the solution brought upon mankind "a deeper, more inward, more poisonous suffering: it placed all suffering under the perspective of guilt." [19]

Although Nietzsche never lost sight of the positive function of asceticism, he expressed profound anger over its abuses and excesses among the ascetic priests who made suffering into a way of life. In *Thus Spake Zarathustra* (1883) Nietzsche began his criticism of the "Afterworldly"—those who long for the rewards of another world. The sick came to despise this world and their own bodies, and in order to escape from the corruption of their earthly existence they condemned their bodies and the pleasures of this world. Zarathustra implores his listeners rather to listen to the voice of the healthy body, which is more honest and purer than that of the sickly. The "despisers of the body" are corrupted by their asceticism and are not to be heeded: "There is more reason in your body than in your best wisdom."

When Zarathustra considers the class of priests who preach death and self-laceration, some of his ambivalence about asceticism emerges. The priests have suffered too much, and therefore they want to make others suffer. Their suffering has made them the most poisonous enemies, such that even those who attack them can be soiled by their vengeful humility. "Yet," Zarathustra confesses, "my blood is related to theirs. . . . I am moved by compassion for these priests. I also find them repulsive." [20] In the last section Nietzsche combined his theory of eternal recurrence with his rejection of asceticism. The complete individual can will at any moment that that moment, being so full, recur eternally. Those who suffer hope for future pleasures of heavenly rewards or long for heirs and for children. But those who affirm their lives, who experience the joy of existence, do not want heirs. "Joy wants itself, wants eternity, wants recurrence, wants everything eternally the same." The overman is able to transcend the endless striving after future joys and find them in his own present life.

In *The Gay Science* and again in *Zarathustra* Nietzsche announced the death of God. He attempted to bury that God in *Beyond Good and Evil*, but in *The Genealogy of Morals* he sought to desecrate His grave. Here his critique of asceticism reached a white heat of intensity.

Nietzsche had previously argued, as I noted, that asceticism was a necessary intermediate stage in the development of the mind from hedonism and its ethic of nihilism to the creative and meaningful life of the overman. In *The Genealogy* he attempted to trace its historical evolution as the dominant morality of Jews and Christians. Hence his evaluation of the accomplishments of the Jews is as ambivalent as his evaluation of asceticism. And though he still noted some of the achievements that asceticism had produced, his condemnation was most severe.

Nietzsche moved back and forth between positive and negative judgments of the "achievement" of the ascetic priest. He was critical of "the anti-sensual metaphysics of the priests" that encouraged fasting and sexual abstinence, but he then conceded that "only on this soil, the precarious soil of priestly existence, has man been able to develop into an interesting creature; . . . only here has the human mind grown both profound and evil." [21] His final judgment of the terrifying renunciation of life produced by asceticism was passionate and absolute. The older morality of a chivalrous and aristocratic nobility revered the virtues of strength, courage, and health. The dominated slave class began to will the misery of its own submissive suffering as a positive value, and so began the slave revolt in morals, one which Nietzsche attributed to the Jews. "It was the Jew who, with frightening consistency, dared to invert the aristocratic value equations good/noble/powerful/beautiful/happy/favored-of-the-gods and maintain, with the furious hatred of the underprivileged and impotent, that 'only the poor, the powerless, are good; only the suffering, sick, and ugly, truly blessed.' " [22] Nietzsche warned of the conspiracy of the sufferers against the happy, the strong, and the beautiful. His argument that "the sick must not be allowed to contaminate the healthy" referred not to unfortunates smitten with disease, but to the sickliness of the masses, whose action tended to deaden the life-affirming and creative energies of the potential overman. "Let us have fresh air, and at any rate get far away from all lunatic asylums and nursing homes of culture!" [23]

Man must learn to counter the excessive guilt and bad conscience that the ascetic ideal has generated, and for that a new nobility is needed. Zarathustra had announced the guidelines for that new class of people who are to overcome themselves and therewith the aesthetic and moral sensibilities of the masses: They must learn to avoid the mediocrity of the herd and its ethic of nihilism and asceticism. They must learn to exercise their will to power, to resist the easy path of conventionality and create a new moral order for themselves. They must be able to "give style to their existence" and ultimately to affirm it wholeheartedly. And that existence will involve a joyous acceptance of the human body. Zarathustra counsels his followers: "Lead back to the earth the virtue that flew away, as I do—back to the body, back to life, that it may give the earth a meaning, a human meaning." [24]

The figure of the dancer in Nietzsche's work is a symbol of the highest achievement of man—a harmonious union of physical strength and rational discipline. "Zarathustra the dancer" would teach his followers to laugh and to dance. For the Greeks the satyr was an expression of man's highest aspirations. He combined in his person the roles of "musician, poet, dancer, and visionary." In the dancer, the will to power has been marshalled over many years to discipline the haphazard and instinctive movements of the body and to create the control necessary to make artful and graceful moving forms. The dancer has transcended hedonism, nihilism, and asceticism, though his ultimate achievement incorporates them in a higher synthesis. His life is truly noble, because he has learned to give style to his existence. He has accepted his corporeality and made of it something poetic and, ultimately, meaningful.

The views of Whitman, Zola, and Nietzsche on human corporeality varied a great deal, but they shared the idea that the sensuous nature of existence must be understood and cultivated. Fear of the body, they agreed, led to various wrong turns in the history of morality, and their works were intended to assist the return to a more rewarding sexual ethic. Whitman came closest to advocating an unchecked hedonism, but he shared with Nietzsche an appreciation of the need for discipline in the service of creativity.

Zola was truly awed by the power of sexuality, and although he tended to imagine situations in which that power was out of control, he followed its destructive consequences in his novels with unmistakable admiration. The work of these three major intellectual figures reveals a persistent interest in the role of the body in the determination of human affairs, an interest symptomatic of a general cultural revision of body image that was under way in their time.

Chapter 9

Pure Women and Superb Men

While physiologists and psychologists explored the ways the human body influenced mental activity in general, a variety of theories was suggested to explain how anatomy influenced personality differences between the sexes. The activity of the feminists in the late nineteenth century added to the interest in "natural" gender roles and gave the literature a strongly polemical tone. The feminist demands for the vote, higher education, and emancipation from the home implied an underlying demand for revision of traditional sex roles. Of particular historical importance were the discussions of women's experiences during the sex act. At mid-century the medical community generally held that active female sexuality is abnormal, but after around 1870 some few daring theoreticians began to view the female orgasm as a desirable consequence of sexual activity. At the same time the medical community began to show a growing preoccupation with the causes and treatment of psychic impotence. The literature on female orgasms and male impotence is an ideological focal point for the changes in male-female sexual relations that were taking place in the latter decades of the century.

Though theories about sex roles included national, class, and ideological factors, a large body of literature emerged which derived male and female character from physical differences. These studies drew extensively from work in evolutionary theory, physiological chemistry, anthropometry, and anatomy.

Herbert Spencer developed an evolutionary theory of gender character which emphasized the more primitive nature of the female personality. Women represent "a somewhat earlier-arrest of individual evolution" which was requisite to enable them to preserve their energy for reproduction. Men are able to devote their energies to intellectual tasks which take them into the world, while women must conserve their energy to bear and nurse children. Since women are physically weaker, they must cultivate attractive qualities to enable them to compete for a sexual partner. Subsequent derivations of female personality from evolutionary theory followed Spencer's argument that the higher evolution of mankind required that males and females cultivate those character traits which were in the service of the species. Spencer also speculated how metabolic rates in men and women influenced character. Women consume less food than men and therefore generate less energy to enable them to perform the taxing intellectual activity which only the male metabolism can sustain.[1]

The Scottish biologist Patrick Geddes also characterized men and women according to their respective metabolic rates and sexual functions. Men expend great quantities of energy to deliver a small cell to its destination, while women conserve their energy to enable them to carry out their reproductive function.[2] These respective metabolic rates lead to the cultivation of specific sex roles. Men excel at high-energy mental activity, while women are temperamentally suited to routine, low-energy domestic activities.

Another group of theories that derived character from anatomy focused on the revelations of anthropometry—a popular scientific study of the proportions of the human body. A number of German researchers compiled statistics to support their theory that a man's brain was larger than a woman's.[3] This fact was used as the scientific foundation for many anti-feminist tracts, such as Paul Möbius's notorious *On the Physiological Feeblemindedness of Women* (1901). Their smaller brains, Möbius believed, make women moody, nervous, and animal-like. They are unable to create, and lying and exaggeration come easily to them. Their mental deficiency, however, prepares them for important functions in the family. They should normally be content with living exclusively for man and child. In accord with the Italian criminologist Cesare Lombroso, Möbius believed that female fertility was inversely proportional to intellect.[4] Others reasoned that the general bodily structure of women—their delicate musculature and nervous system, their frailty—fated them to nervous irritability and emotional weakness.

Some studies viewed the breast as the essence of womanhood. Jules Michelet's influential *L'Amour* (1858) pointed out that a woman's shallow breathing produced the unique feminine undulation of the bosom "which expresses all her sentiments in a mute eloquence."[5] A German critic, Leo Berg, speculated that the breast alone determined the entire personality of a woman. "A woman's breast is the organ with which she is able to express herself most intelligently. It is her language and poetry, her history and her music, her purity and her desire. . . . The history of the corset and the body is almost the whole story of the female sex. The bosom is the central organ of all female ideas, wishes, and moods."[6] The medical literature was particularly incensed about the use of ointments for enlarging, shrinking, or preserving the firmness of breasts. The identification of feminine virtue with childbearing and nursing forbade such tampering with the breasts for cosmetic reasons. One mid-century physician eulogized "the caucasian female bust" as "the crowning act, the most elaborate and most perfect form and model of beauty, human eyes have yet beheld."[7] In the late 1890s the "bosom ring" came into fashion briefly and sold in expensive Parisian jewelry shops. These *"anneaux de sein"* were inserted through the nipple, and some women wore one on either side linked with a delicate chain. The rings enlarged the breasts and kept them in a state of constant excitation.[8] This provocative ornamentation was rare, but a larger number of women took to padding the breasts, and some had pieces of rubber inserted under the skin. The medical community was outraged by these cosmetic procedures, for they represented a rejection of traditional conceptions of the purpose of a woman's body.

A recurring article of popular belief was that menstruation was the key to the female temperament. This chronically bleeding wound was a sign of her unfinished state—a cyclic incapacitation which drained her vital energy and rendered her incapable of handling responsibilities that required uninterrupted health. In the first half of the century medical authorities believed that women, like animals, were fertile during menstruation. This presumption no doubt foiled many couples trying to work out a rhythm method of birth control. In 1854 the renowned expert on contraception, George Drysdale, could still write that "menstruation in woman corresponds exactly with the period of heat in female animals." [9] Throughout the century menstruation was viewed as a kind of disease. Michelet concluded that women are "invalids" because of their menstrual cycle. His famous quip *"La femme est une maladie"* was often repeated in misogynist tracts.[10] Even the enlightened Henry Maudsley added to the corpus of superstition by charging that women were unfit for responsible professions because for one quarter of each month they were "more or less sick and unfit for hard work." [11]

In 1905 the German feminist Rosa Mayreder published a critique of some theories that had been suggested to explain female character from the female body. Möbius and Lombroso argued that women were physically and psychologically convervative; the Goncourt brothers concluded that women tended to excess in everything; Bachofen found them dominated by instinct; and Laura Marholm spoke of woman's natural flightiness and unreliability. Woman's body, others noted, was soft, and therefore her mental capacity was unsuited for precise analytical work; women relied on their natural intuition, which figured so prominently in their personalities. "According to the fundamental assumptions of modern natural science," Mayreder wrote, "every manifestation of consciousness is linked with a bodily process." She cited an example of this reasoning from the eminent German biologist Rudolf Virchow, who maintained that "all the characteristics of woman's body and mind—nutrition and nervous activity, the sweet tenderness and curves of her limbs, the development of her bosom, her posture and voice—in short everything that we revere in women is dependent on the activity of her ovaries." [12] Mayreder rejected this kind of sexual-biological determinism. Rather, she said, these characteristics were culturally determined, and a more useful sexual psychology would have to steer between the extremes of biological determinism and the equally exaggerated claim that women were in every respect equal to men. Mayreder offered further insight into this theorizing about femininity when she charged that men preferred a frigid woman to one as sensuous as themselves. This speculation contributed to a subject that attracted ever bolder analyses toward the end of the century as the stereotypical roles of male domination and female submission were questioned.

The literature on female sexuality addressed itself to four fundamental questions: (1) Do normal women experience sexual excitation? (2) Ought they to show pleasure during the sex act? (3) Do women have orgasms? (4) Do men or women have the greater capacity for sexual activity or for sexual pleasure? There was a steady rise in the frequency with which these questions were openly discussed in the latter part of the nineteenth century, and the first three questions were more and more consistently answered affirmatively. The debate over the capacity and intensity of male versus female sexuality remained a lively one, and some daring men began to acknowledge that women did have the greater sexual capacity.[13]

The most often cited authority on female sexuality in England was William Acton, whose treatise of 1857 insisted that normal women did not have a strong sexual instinct. "As a general rule," he wrote, "a modest woman seldom desires any sexual gratification for herself. She submits to her husband, but only to please him; and, but for the desire of maternity, would far rather be relieved from his attentions." [14] One mid-century marriage manual warned that "voluptuous spasms" could make a woman barren. A study of 1880 concluded that only one woman in four even knew that she had a clitoris, because it was insufficiently stimulated during coitus.[15] In 1878 George Napheys dismissed the "vulgar opinion" that women "are creatures of like passions with ourselves." He concluded that "only in rare instances do women experience one-tenth of the sexual feeling which is familiar to most men." [16] Another moralist advised that the mucous fluid that was sometimes secreted during intercourse "only happens in lascivious women, or such as live luxuriously." [17] As late as 1894 a German gynecologist argued that satisfaction of a woman's sexual needs lowered her life expectancy.[18] The medical literature on female sexuality shifted between insisting on the lack of sexual instincts in women and moralizing on behalf of preserving the "inborn prudery" of respectable women.

To maintain the ideal of female chastity and restrict women to a passive sexual role in marriage required great pressure to keep women from acknowledging or indulging their sexual impulses. Most of the medical literature helped impose the idea that the male sexual impulse was natural and acceptable, while the female sexual impulse was unnatural or even pathological. Women were warned about the multifarious evil consequences that could result from too early an awakening of their sexual impulse or an exaggerated cultivation of it. Some physicians warned that the sexually active woman could inhibit her husband and make him impotent. Others warned that they could not conceive children if they moved during the sex act. Physicians offered a host of confusing and threatening warnings about what might happen to their internal organs, their emotions, their offspring, or even their husbands as a consequence of any overt manifestation of their sexuality.

There is some evidence, however, that beginning in the 1870s ideas started to change. One historian has concluded that "although most doctors had denied or remained silent about the female orgasm, after 1870 it began to get objective consideration by a number of physicians." [19] One strongly moralist tract of 1881 shows the growing willingness to discuss the female orgasm. In spite of the author's repeated warnings about the dangers of masturbation, he explicitly condoned female orgasms provided they occurred during the "conjugal act" and not by masturbation. "The orgasm induced in the female organs by the conjugal act is such that, if left incomplete, the congestion does not immediately relieve itself, and inflammation, ulcerations, and final sterility are the results." [20] This guarded concession to the legitimacy of female sexuality constituted a substantial shift from the negative pronouncements on the subject made by Acton in 1857.

The most compelling evidence for female sexuality was female masturbation, a practice which even the boldest advocates of liberal sex ethics usually avoided. One moralist referred to female onanism as a subject "too horrible to contemplate," although he then proceeded to contemplate it at some length. That mixture of indignation and voyeurism was characteristic of much nineteenth-century medical literature.

In addition to the standard list of possible causes of masturbation among young boys, a few additional sources of temptation were thought to lead a young girl to practice the "solitary vice." A German nudist spokesman reported on a sermon in a Frankfurt newspaper in which women were warned not to hold a coffee grinder between their legs for fear that the vibration might lead to sinful stimulation.[21] Havelock Ellis explained that some women used the vibratory motion of a railway train to masturbate. Others became excited by bicycling.[22] Ellis attracted considerable attention with an account of another possible cause of masturbation. While touring a clothing factory he observed a number of women who began furiously working the pedal of their sewing machine while their eyes took on a distant stare and finally closed in the pleasure of a sexual orgasm. Ellis reacted with disgust in this report that no doubt horrified many of his readers. The image of a young girl rubbing her thighs together at a sewing machine or manually stimulating herself shattered the traditional ideal of the sexually passive and demure woman. Consider one characteristic expression of fear and outrage by an American physician in 1881 on the subject of "female self-abuse."

> Alas, that such a term is possible! O, that it were as infrequent as it is monstrous. . . . We beseech, in advance, that every young creature into whose hands this book may fall if she be yet pure and innocent, will at least pass over this Chapter, that she may still believe in the chastity of her own sex; that she may not know the depths of degradation into which it is possible to fall.[23]

The author of this tantalizing warning goes on to list the numerous diseases that masturbation can cause—similar to the horror show envisioned for the masturbating male, though in addition women were subject to an excessive vaginal discharge ("the whites") and sexual disturbances including nymphomania, epilepsy, and hysteria. Treatments for female masturbation included the bland diets, censored reading, and threats prescribed for boys, but the extreme measures far surpassed the most sadistic tactics used on young men.

The ultimate masculine assault on female sexuality—surgical removal of the clitoris—is one of the most tragic chapters in the history of medicine. The operation was developed sometime in the early nineteenth century. It was first used in Berlin in 1822.[24] It was popularized by a London surgeon, Isaac Baker Brown (later to become President of the Medical Society of London), who developed the operation of the clitoridectomy in 1858 to treat cases of what he judged to be excessive masturbation. The operation has a sporadic history—sometimes recommended, sometimes violently criticized—but was used intermittently until sometime in the 1920s, when it was performed for the last time in the United States.

A French medical encyclopedia explained the prevailing view of clitoridectomy in 1881. "This operation has attracted much attention during the past few years, but although it has produced some favorable results when employed by conscientious surgeons, today it is generally not used. There is just too much evidence to show that the clitoris does not play either an exclusive or a preponderant role in orgasms." [25] The popular French medical writer Pierre Garnier surveyed a variety of treatments for female masturbation in a treatise of 1883. He argued that the *"foyer clitoridien"* scarcely provoked any excitation unless it was abnormally enlarged from habitual masturbation. If cauterization of the clitoris did not work to quell excessive masturbation then, he concluded, the clitoridectomy might be used as a last resort. Garnier assured his readers that the fear of some physicians that they might deprive their patients of all future voluptuous sensations was entirely unfounded. "This error is demonstrated today by the numerous cases of women who have had this organ removed." [26]

In 1884 the German neurologist Paul Flechsig recommended castration (removal of the ovaries) in the treatment of female hysteria and discussed several cases in which he had employed it.[27] The German gynecologist Alfred Hegar reported on castration treatment of hysteria in 1885. Generalizing from his therapeutic success with castration, he argued that sexual substances affected the entire nervous system.[28] The French physician Godefroy Thermes reported on the effectiveness of clitoridectomy operations by two of his colleagues. One performed sixty-two operations, from which there resulted eight deaths and twenty-five "cures." Another made twenty-two clitoridectomies with ten reported cures. Deferring to the many physicians who had renounced castration in the treatment of hysteria, Thermes concluded that it ought to be performed only when organic disorder of the sex organs was certain.

The emergence of women's sexual consciousness was greeted by a full spectrum of negative responses ranging from the fear and recoiling of sexual impotence to the sadistic surgical counterattack. Most men adjusted with a minimum of trauma, but the extreme reactions illustrate the magnitude of the changes that were taking place. While some literature on gender roles in the late nineteenth century began to consider the advantages of acknowledging, or even cultivating, active female sex roles, a large body of material reveals the traditional man defensive and protesting. One reaction to emergent female sexuality by the men was sexual confusion and sexual failure. The male sexual ego, conditioned as it was to always dominate passive female sexual objects, was for a time paralyzed by the new sex roles that accompanied the feminist movement. Of course not all men became impotent, but the literature on manliness reveals an increasingly anxious preoccupation with the causes and treatment of psychic impotence.

A good deal of the literature expressed concern that modern men were becoming weaker from the introduction of labor-saving machines. One advocate of a program of physical exercise argued: "It is only necessary to visit a Turkish bath to find abundant evidence of the muscular collapse which has overtaken the modern city-dweller: bodies 'developed' everywhere in the wrong direction; arms like pipe stems, spindle thighs and straight calves weakly support bellies like Bacchus." [29] A pioneer in the physical culture movement in America, Dudley Allen Sargent, worked out a series of exercises which were intended to imitate the muscular movements made in traditional kinds of labor. Members of his gymnasium in the 1880s were instructed in exercises which resembled wood-chopping, rowing, and hammering.[30]

The manuals on body building that appeared in the 1890s reveal the insecurities that plagued some men conscious of their changing sex roles. William Blaikie's *How to Get Strong and Stay So* (1898) surveyed the causes of the degeneration of the male physique from the reduction of natural physical labor, advised the reader on a series of exercises to restore manly strength, and concluded with a two-hundred-page history of "Great Men's Bodies" from ancient times to the present.

One of the big attractions of the circuses which enjoyed so much success in the 1890s were the various strong-man acts. Tom Burrows dazzled the British public by swinging an Indian Club for 8¼ hours.[31] The famous French Hercules "Apollon" was exhibited often in France between 1900 and 1914.[32] One historian of the circus viewed the success of the German "Rasso Trio" in the 1890s as part of a general interest in physical culture.[33] A cult of the strong man formed around the person of the American physical culture enthusiast Bernarr Macfadden, who outlined some of the most desirable masculine traits in *The Virile Powers of Superb Manhood* (1900). He offered his program of exercise "to all those whose souls and bodies are tortured by weakness." The book, he explained, would help men to be men—"strong, virile, superb." He challenged his readers by insisting that if they did not possess "true manhood," their first duty was to acquire it. "If you are not a man," he threatened, "you are nothing but a nonentity!"[34] Macfadden assumed a close relation between physical health, a muscular physique, and sexual potency, and underlying the positive counseling was the terrifying threat of sexual impotence.

Eduard Fuchs speculated about some of the social and class origins of the masculine ideal, but also came to focus on the fully potent husband and father as its apotheosis. The complete man should be energetic, strong-willed, self-confident, with an upright posture and a firm grip. These characteristics also implied that he be sexually energetic, strong-willed, and confident—constant, upright, and firm. His sexuality must not be squandered, but postponed for future gains and channeled into productive activity.[35]

One uniquely nineteenth century fear was a disease called "spermatorrhea"—involuntary discharge of semen with or without an erection either during sleep or waking hours. The disease was first described by the French surgeon Claude-François Lallemand in 1836 and came to the attention of a number of physicians in Europe and America during the latter decades of the century.[36] The disease was believed to be a corporal punishment for the squandering of sexual energy, and the various treatments illustrated the fear it generated. The most elaborate devices were designed to awaken the sleeper when he got an erection and thereby avert a nocturnal emission. Metal rings were used to apply pressure to the expanding erection, and J. L. Milton described a gadget which awakened young men with an electric bell: "a ring placed on the penis is so made, that when expanded by erection it completes an electric circuit and so rings a small alarm bell placed under the sleeper's pillow" (fig. 19).[37] John S. Haller and Robin M. Haller have unearthed one extraordinary treatment for spermatorrhea. According to the Hallers, the French physician Armand Trousseau "recommended the insertion into the rectum of a wooden ovoid cylinder the size of a pigeon's egg, large enough to compress the prostate gland." The egg was believed to redirect emission backwards into the bladder.[38]

The various humiliating treatments for this imaginary disease suggest that misunderstanding of the normal functioning of the sexual apparatus generated a great deal of panic among some nineteenth-century men. Our knowledge of these procedures is limited largely to the medical literature in which they were recommended, and historians know even less about the extent of public compliance. It appears to be rather certain, however, that there was a well-understood central message behind all the counseling and threatening about masturbation, nocturnal emissions, spermatorrhoea, and excessive intercourse—that ultimately they led to the most dreaded functional disorder of all—psychic impotence.

Pierre Garnier's five-hundred-page study of 1881 offered French readers a detailed analysis of the causes and treatment of impotence. A major concern of the French at that time was the apparent decline in the birth rate, and, by way of explanation, Garnier speculated that "the majority of Frenchmen today are suffering from infertility of epidemic proportions!" [39] Most of the causes of impotence he found were of a physical nature—obesity, venereal disease, nervous disorders, abnormally large or small sex organs, and the like. He also discussed various toxic substances that troubled men in certain professions—mercury poisoning among mirror workers, carbon sulfide among rubber workers, lead poisoning among miners, and arsenic used in a variety of industries.

His evaluation of the effect of women's sexual performance on male potency was equivocal and revealed the transitional nature of his work. On the one hand, he argued that women must take some active role; total immobility of a woman could contribute to a man's failure. Some men, he warned, interpreted woman's "natural prudishness and reserve as indifference and frigidity." But his general assessment was in accord with the conventional view that women should have a lesser physical involvement than men. The woman, he argued, "tastes the surplus voluptuous feelings more in her heart and soul. The normal role of women during copulation is much more moral than physical. Otherwise it would only be an unequal battle between unequal organs." [40]

Garnier concluded that the responsibility for initiating sexual activity must be left entirely to the man, though during the sex act women ought to move and experience pleasure in moderation. Although Garnier acknowledged the legitimacy of female sexual desires, he was particularly concerned about the fulfillment of male sexual needs and cautioned that the overly aggressive woman might diminish his pleasure. One American physician expressed this fear explicitly in a sex manual of 1881: "A strongly passionate woman may ruin a man of feebler sexual organization. . . . Whatever may be her feelings, she should always remember that delicacy [requires] her to await the advances of her companion before she manifests her willingness for his approaches." [41] The revealing phrase in this prescription is "whatever may be her feelings." The author acknowledges that women will have sexual desires, and he does not condemn them so long as they do not interfere with the man's. Thirty years earlier it was far less likely that such a book would mention female sexual desires at all without being strongly critical of them.

In 1896 a German physician introduced his study of male impotence with the complaint that the subject had been neglected among medical circles and confused by hypocrisy. He began with a list of causes of impotence which must have intensified his readers' insecurities. Small genitals, he argued, were always a sign of "insignificant" sexual powers. Some large genitals, he acknowledged, did not fill with blood properly during erection, but this was caused by excessive pre-pubertal masturbation. To complete his intimidation of his fair-skinned German readers he added: "Stronger pigmentation in the sexual organs is generally accompanied by greater capacity in venery, which is seen in negroes, who are, as a rule, endowed with large genitals." [42] Among the many listed causes of impotence was sexual excess, which, he explained, was a particular problem at that time because men felt the need to impress women with their enormous sexual capacity; this practice was intensified by "voluptuous women," the new generation of "over-sensual" women who were challenging male sexual ability. The treatment he recommended concentrated on the need to lessen the man's anxiety over failures, distract him from the problem, avoid useless sexual excitation, and follow a routine of gymnastic exercises. He reluctantly allowed the use of one apparatus, the "Sledge" (two splints connected at the base with a metal ring and at the upper end by a soft rubber ring), to artificially sustain an inadequate erection.

Some of the mechanical treatments for impotence reveal the pathetic extremes to which the Victorian man went to rid himself of this deficiency. The Hallers have described "electric belts," "rectum medication," and "crayons inserted in the urethra" employed by some men in the hope of restoring their potency.[43] One American physician inserted one electrode into the rectum, another into the urethra, and a third between the penis and the scrotum or on the inner side of the thigh.[44] These galvanic therapies were intended to restore the lost electrical or magnetic energy which it was believed helped drive the male sexual apparatus.

In 1908 the German physician Emil Peters speculated that impotence was a frequent cause of marital unhappiness and many unexplained suicides. The rearrangement of sexual relations consequent to developments in the feminist movement added to the incidence of impotence at that time. In accord with most experts on the subject Peters warned about the dangers from sexually aggressive women. "A lively and passionate woman can severely damage a man, causing impotence."[45] The irrepressible sexual moralist Sylvanus Stall offered some advice to impotent husbands. A few virtuous men, he conceded, may have become distraught from the excitement of marital relations, but, he assured his readers, a few months' exercise with Indian clubs would more than likely remove their fears and restore their masculinity.[46] One can hardly conjure up a more pathetic image than one of Sylvanus Stall's young men, isolated and terrified by his sexual impotence, trying to restore his potency by inscribing exercise patterns in the air with Indian clubs.

The evidence available from the medical literature allows a few speculations about the revision of male-female relations in the late Victorian period. There is no way to prove that the incidence of impotence was higher than in any previous period, nor that it was more widely discussed, but the nature of the evidence—the ominous warnings and the bizarre treatments—suggests that men viewed the problem as particularly threatening at that time. Most men can handle an occasional failure, but when they believe that the conditions which created it will persist, then they may be thrown into a panic which can itself generate further impotence. A significant number of men at this time perceived a connection between their occasional failures and the changes in sexual relationships that were occurring in conjunction with the activity of the feminists. The sexual aggressiveness of women was not always consciously articulated or even consciously acted out, but it was implicit in the logic of the feminist movement, which was very real and very threatening to many men. Though it is impossible to know for certain whether the pattern of male-female sexual relations did undergo significant change at this time, the evidence shows that it was believed to be changing. The several warnings that a lively and sensuous woman could damage a man by rendering him impotent are telltale signs that those changes were taking place in the popular imagination.

Chapter 10

Physiology of the Victorian Family

In the course of the nineteenth century the emotional bonds within the European and American family grew steadily stronger, and by the end of the century both parents and children began to seek refuge from the forces that held it together with such overbearing intensity. Members of a family were believed to be linked by bodily fluids and vaporous exhalations which exerted far-reaching influences on their physical and mental development. Victorian medical literature repeated the ancient idea that the exchange of sexual substances between parents was essential to the well-being of both, especially to the wife, who would be subject to hysteria if her sex organs were not regularly bathed by the soothing balm of semen.[1] The medical literature on the parent-child relationship argued that the entire emotional history of the parents would somehow be transmitted to the child. The physical and mental condition of the parent during conception was of paramount importance, and, following conception, it was believed, the slightest shock or unpleasant thought would affect the future child. Pregnant women were frequently advised to refrain from sex altogether. If they did not, the child might develop some sexual anomaly; if they drank, the child might become an alcoholic; and if they experienced any intense emotions, the child might be demented. Of course the mother's responsibility for the mind and body of the child was believed to be the greater, but even the father's bad habits before his marriage, his youthful "excesses," and his slightest deviations from the strictest propriety would be imprinted indelibly on his children.

The physical relationship between the parents was shaped by the faulty sex education and limited sexual experience of both before marriage. Consider what the typical middle-class Victorian couple might have known about each other's bodies on their wedding day.

Since adolescence the man has been swamped with threats about the evil consequences of masturbation, and he has grown to view sex generally as dangerous to mind, body, and morals. He has been led to believe that any sexual discharge, even during marital intercourse, will sap his energy. If he has had any pre-marital sexual relations, they have probably been with a prostitute or a servant girl, and he has therefore become confused by class prejudices involving his role as the dominant partner. His experience with prostitutes has ill-prepared him to relate to his virginal wife. If he has taken his chances, he may be plagued with the suspicion that he contracted a venereal disease which could be communicated to his bride or future offspring. During his courtship he has had no opportunity to see his fiancée's body, and of course he has no idea of how she might respond sexually. He has seen only her face and hands in the flesh, though perhaps he has had a glimpse of her bosom pushed out of a corset under a low-cut evening dress. He does not know what her legs or torso look like, and, since he has never seen her doing any vigorous athletic activity, probably has little idea of how her body moves freely, unencumbered by a welter of clothing. There has been no discussion of sex between them: menstruation is an unmentionable secret, her previous sex life is presumed to be non-existent, and her expectations about her future sex life are vague and confused. Their physical contact has been limited to an occasional embrace during a waltz, formal kisses on the hand, and perhaps a daring touch of their knees under a table. Neither has any idea of how the other smells—all natural body odors have been obliterated with perfume, talcum powder, and scented pomades.

If he has read a marriage manual, it has probably explained that if his wife is normal she will not manifest any strong sexual desires and will await his advances and receive them passively. Perhaps he has received some counseling about the proper frequency of engaging in sexual relations. If he has read Sylvanus Stall, he has learned that they should have sex once a week and never undress in front of each other in order to avoid excessive stimulation of the sexual appetite. (In 1911 the Italian sociologist Robert Michels could still complain that European couples were ashamed to undress before each other and had to put out the lights before having sex. He argued that women were often ashamed of their own sexual feelings and therefore hid and disguised them. Some women would develop a false modesty which confused the man, for whom it would become impossible to distinguish between flirtatious coquetry and genuine resistance. Other women, Michels noted, would carry out their posturings of resistance so thoroughly that the man would be forced to be brutish.[2] To avoid sadistic excess, young men were cautioned by manuals of the time not to be too forceful with their brides. One doctor reported in 1889 that he had treated one hundred and fifty women for injuries sustained during their wedding night.[3])

The young groom might discover during the honeymoon that his bride was menstruating. But he could hardly blame her, because the social calendar probably dictated the date of the wedding, and there were few families in which a young girl could comfortably discuss such matters with her parents, let alone with her prospective husband.

Contemporary sex manuals universally counseled against intercourse during menses. Some said that men could become diseased by having intercourse with a woman during her "turns." George Napheys argued that a young man could contract gonorrhea from such a sexual contact. Another German physician rejected that theory but expressed strong disapproval of any sexual contact while a woman is "unclean."[4] The various fears and superstitions about the menstruating woman that persisted throughout most of the nineteenth century further complicated the search for mutually gratifying sex roles.[5]

The woman's body was viewed as both a haven of pleasure and a vessel of sin. Her entire being was believed to be intimately linked with her sexual functions—particularly menstruation, childbearing, and nursing—but at the same time she had to deny the intensity of her sexual instincts. She was not to enjoy, or at least not to show, sexual pleasure, but she was to receive a man's sexual embrace as though it were the ultimate consummation of her physical and spiritual existence. She had to be a virgin, while her husband was expected to have had an extensive apprenticeship. Some desperate women faked their virginity by inserting a capsule filled with chicken blood into their vagina on their wedding night. Others had a physician make a few stiches in their vagina which would rip open and bleed during their "defloration." [6]

Conflict between men and women derived in part from the estrangement that the conventional morality imposed on them. One extraordinary anecdote illustrates how complete that estrangement could be. A German widow said that she noticed that her husband had blue eyes only when she saw him lying in his coffin. "As long as he lived I would never have dared to look at him so closely." [7] The confusion generated by fear and ignorance about the relationships between parents was matched by an even greater confusion about the relationship between parents and children.

GERMS, GENES, AND IDEAS

In the latter half of the century the educated classes began to acquire bits and pieces of the discoveries in physiology that were being made with such impressive consistency. But the information that filtered into the marriage manuals and handbooks for the young was often worse than none at all. Confusion about the nature of cultural, infectious, and hereditary transmission from parent to child added to the intensity of the pyschological bonds between members of a family. Accordingly, it was believed that if a child's parent was insane, tubercular, or syphilitic, the child could inherit those problems. The nature of the three ways a parent could transmit his germs, genes, or ideas to a child was not precisely understood even in the higher circles of the medical world, and a bewildering tangle of ideas plagued the general population. The confusion tended to produce a conception of the family as a biological unit in which the values, sex cells, and germs of parents worked in and on the minds and bodies of children, leaving little room for individual development. Children came to feel trapped in their families, fated by the flow of mysterious substances through their bodies to repeat the same vices, catch the same diseases, and inherit the congenital weaknesses of their parents.

Weismann's theory of the immutability of the sex cells only began to appear in popular literature at the end of the century. For the most part Lamarck's theory of the hereditary transmission of acquired characteristics was accepted. This view of heredity was elaborated by many others who frequently magnified the parents' sense of responsibility for their offspring. The American essayist Georgiana Kirby explained that the physical and mental experiences of the mother during conception and gestation determined the character of the child. According to Kirby, the moment of conception was most important: "Never run the risk of conception when you are sick or over-tired or unhappy. . . . For the bodily condition of the child, its vigor and magnetic qualities, are much affected by conditions ruling this great moment." [8] She also claimed that unpleasant experiences and thoughts could be communicated to the fetus during pregnancy. Most marriage manuals argued that a pregnant woman ought not to have sexual relations for fear that the developing embryo might be adversely affected. The magical power attributed to semen for energizing the man's psychic apparatus was believed to have a negative effect on the unborn child. Moralists speculated that such a violent interruption of the intrauterine experience might incline the future child to have aberrant sexual impulses.[9] A French marriage manual even advised the pregnant woman to keep other people away from her room and insist that whoever entered be scrupulously clean to avoid contamination of her future child.[10]

Another bewildering line of reasoning about the intimate biological relationship between fetus and mother held that the blood of the child's father became mixed with that of the mother because she shared her blood with the fetus. An American physician argued that "the child has impregnated the mother with the blood of the father. Successive pregnancies can only add to this admixture." [11] Eduard von Hartmann (1895) argued that if the mother conceived a child with a second man, the child would show the influence of her first husband. "The husband of a widow does not therefore find a clean page, but one written over by his predecessor, with whose hereditary tendencies his own must enter into conflict." [12] In 1903 the German essayist Otto Weininger expounded the same theory with a racist twist. "White women who have borne children to black men, and who then bear children with white men, are said to have retained enough of an impression from the first mate to show an effect on the subsequent children." [13] Strindberg dramatized this theory in *The Creditors* (1888) by having a woman's first husband return to haunt, and eventually destroy, her second marriage. Her second husband explains how "when the child was three years old it began to look like him, her former husband."

The German physicians Wilhelm Fliess and Hermann Swoboda believed that this transmission followed the rhythm of the cyclic movement of all life. In 1897 Fliess explained that the mother transmitted her periodicity to her child. In his vast study *Der Ablauf des Lebens* (1906) he published hundreds of tables illustrating a periodic unity in family life, in which attacks of enuresis and diarrhea, backache and headache, appeared in the mother and her children in combinations of twenty-three- and twenty-eight-day cycles. Swoboda expressed the same idea that "the organism is subject to a rhythmic flux" transmitted to the child at birth.[14]

Samuel Butler's response to his own family experience alternately expressed two elements of a strong ambivalence. In his earlier studies of the psychological and biological mechanisms operative within the family in *Unconscious Memory* (1880) he described "the oneness of personality between parents and offspring" [15] without objecting to the intimate link between parent and child, whereas in *The Way of All Flesh* (1903) he reacted violently against the threatened imposition of that "oneness" between himself and his father. Many shared his later view that the tightly knit family imposed itself excessively on the individuals in it.

The most widely discussed aspect of heredity was degeneration—the transmission of destructive properties to the offspring. The lack of an accurate account of sexual reproduction led to the theory that acquired germs and vices remained forever locked into the germinal cells of the individual and accumulated from one generation to the next. A popular source for the many theories of degeneration that circulated at this time was B. A. Morel's *Treatise on Degeneration* (1857). Morel outlined the sequence of four generations in which a family line is finally destroyed.

First generation: nervous temperament and moral depravity.
Second generation: tendency to apoplexy, neurosis, and alcoholism.
Third generation: mental disorders, suicide, and defective intellect.
Fourth generation: congenital idiocy or feeblemindedness, physical malformations and sterility.

Numerous variations on this sequence were suggested, and many attempted to observe its progress from generation to generation. Zola organized his novels about the Rougon-Macquart family according to such a sequence of family decline. Germs and vices as well as "internal secretions" were believed to be at work in the family, undermining health and morality in inscrutable ways. The French educator Gabriel Compayré argued that morbid phenomena were especially subject to the laws of heredity, and accordingly "the bad is transmitted much more easily than the good." He claimed that the biological unity of family life reached over the generations to carry the curse of ancestral sin and "unrestrained acts." "There has been some time—a day, an hour—in which the fate of an entire family has been cast, so that a mutual moral responsibility binds parents to children." [16]

Though much Victorian moralizing about the evil consequences that excessive indulgence of the sexual instinct would have on children was alarmist and biologically unfounded, the problem of venereal disease gave these warnings substance. There were countless family tragedies resulting from untreated cases of syphilis. Chaste women developed syphilis chancres soon after marriage and had numerous sexual problems and miscarriages. They commonly believed themselves to be at fault and explained their problems as some inexplicable "female" reaction to married life. The theme of the father's sins being visited upon his children appeared in two important dramas of the Victorian period.

Without mentioning the name of the disease, Ibsen in his play *Ghosts* (1881) explored some of the fears that syphilis produced. At first Oswald Alving believes that his weakness and loss of will are the consequences of his own debauched living, but in due course he learns that he has inherited the product of his father's recklessness.

"The sins of the fathers shall be visited on the children," he says to his mother, who herself does not know the true cause of his illness. She believes that in some mysterious way the "ghosts" of the past remain alive in children and lead them to repeat endlessly the same life style as the parent. She pairs the physical with the mental heritage as she explains to her son: "It isn't just what we have inherited from our father and mother that walks in us. It is all kinds of dead ideas and all sorts of old and obsolete beliefs." [17]

In 1901 Eugene Brieux published a play about syphilis, *Les Avaries* (Damaged Goods), in which the disease is mentioned explicitly. Before the curtain the stage manager appears and warns the audience that the play concerns syphilis. George is told by his doctor that he must postpone his marriage for four years and undergo treatment for syphilis. Pride and social pressures make it impossible for him to comply. He marries too soon and produces a syphilitic child. In Act II the doctor fights with George's family to get them to dismiss the wet-nurse whose own health is threatened by the diseased child. Brieux points out not only the dangers of syphilis, but the hypocrisy which compels George to marry before he is cured, permits the wet-nurse to endanger her own life and that of other children whom she may nurse, and leaves the innocent mother a hopeless untouchable.

The spread of syphilis was widely believed to be but one of many problems leading to a degeneration of man and society. It was feared that the family was dying out—that, along with the destructive forces of disease and vice, social and economic changes were discouraging the continuation of the large family. Hauptmann's *Before Sunrise* (1889) portrayed the advanced condition of degeneration of the Kreuse family. The children suffer from the parents' indulgence in incest, alcohol, and tobacco. A doctor observes of the family that "there's nothing but drunkenness, gluttony, inbreeding and, in consequence—degeneration along the whole line." The play concludes with the suicide of the most promising child, Helene, who is abandoned by her lover when he discovers that she may inherit her parents' alcoholism. A macabre mother-child relationship is explored in a short story by Thomas Mann, *Tristan* (1902). The mother enters a tuberculosis sanatorium ten months after giving birth to her son. For some unknown reason the child's increase in vigor debilitates the mother. In the end she dies and the child lives. It was also widely believed that parents could rob children of strength. This idea was popularized by William Acton, who argued that great men usually sire weak children, because their vital energies are exhausted by their own lives, which somehow deplete those unknown substances that go into the production of future generations.[18]

THE SCHREBER CASE

One father-son relationship illustrates in the extreme a parent's manipulation of the corporeal existence of a child. I will discuss it in some detail because both the father and the son left records of their ideas and experiences. These records reveal some of the most intense family psychodynamics of the Victorian period.

Daniel Gottlob Schreber founded a program of therapeutic gymnastics in Germany and developed a system of exercises to prevent or correct physical deformities in growing children. His guiding principle was that the body should be maintained in a tensely erect posture *(eine straffe Haltung)* at all times. In 1859 he announced his goal for the proper development of the human body: a symmetry between upper and lower, right and left, to be achieved by means of a regime of compulsory instruction in youth to create a "balanced equilibrium" of the body, which is then maintained in adulthood unconsciously and instinctively. "No part of the body will be neglected in the effort to give all postures and movements a noble tension." [19] Schreber devised a series of exercises to ensure such a symmetrical and rigid physical development; to counter the natural tendencies of children to deviate from his norm, he invented braces to keep the body in form. A "head-holder" was attached to the hair and the underclothing so that whenever a child turned his head, the device would pull his hair. Another contraption held back the child's shoulders. The "upright-holder" was a metal bar that pressed against the chest of a child sitting at a desk and prevented him from leaning forward (fig. 20). Schreber tested these various apparatuses on his own children, and from the memoirs of one of them, Daniel Paul Schreber, the ideas of the father came to the attention of Freud and influenced an important case study.

The career of Daniel Paul Schreber as a judge was interrupted in the 1890s by mental illness. During periods of lucidity he wrote some memoirs which Freud used as source material for working out a theory of paranoia—Schreber's basic symptom pattern.[20] In the 1950s a New York psychoanalyst, William G. Niederland, noticed some connections between the delusions in Schreber's paranoid system and some of the apparatuses and exercises that he had read about in the pamphlets on therapeutic gymnastics written by the father. Niederland reconstructed Schreber's childhood by assuming that the adult paranoid delusions were memories, or slightly disguised memories, of actual childhood experiences.[21]

Schreber's entire body image was affected: "I may say," he wrote, "that hardly a single limb or organ in my body escaped being temporarily damaged by miracles." [22] The purpose of the delusional system was his emasculation and transformation into a woman to equip him to become impregnated by rays from God and breed a superior race of men. One delusion, which Schreber described as the "head-being-tied-together-miracle" *(Kopfzusammenschnürungswunder)*, Niederland traced to his being subjected to wearing the head-holder (fig. 21). His delusion that he had no stomach may have been suggested to him as a child by the presence of drawings and plaster casts of dissected abdomens which his father studied. Some of his father's exercises were designed to prevent the internal organs from developing asymmetrically. The son's "coccyx miracle" was from an exercise to make sitting or lying down impossible. The father had insisted on a special training for sitting: ". . . one must see to it that children always sit straight and on both buttocks simultaneously. Children should be forced to hold themselves upright and erect." [23] Niederland interpreted the central delusion of Schreber's emasculation as a consequence of the generally castrating atmosphere that prevailed in the Schreber home and the castrating effect upon the children of consistently binding, bracing, and disciplining their bodies.

Some of the father's moral injunctions further desexualized the son's emotional life. He believed that all bodily appetites ought to be severely controlled from earliest infancy. A child should not be caressed to comfort him, and feeding should be only for nutritional purposes. All signs of the child's willfulness should be crushed, and corporal punishment should be freely employed. A regimen of exercises should be used to develop self-discipline, and where self-control is lacking, the various apparatuses he devised should be used to keep the body in the proper tensely erect posture. A sense of orderliness and cleanliness should be instilled by early toilet training and the insistence on immaculate body hygiene. Parental discipline should be severe and consistent to achieve complete control. If parents succeed in controlling the child's inner sentiments as well as his outer actions, "they will soon be rewarded by the appearance of a wonderful relationship where the child is nearly always ruled merely by parental eye movements." [24]

The Schreber case has historical significance beyond an account of the individual problems of a man subjected to experimentation by a sadistic father. Schreber's body acted as a spokesman for his age and revealed many of the pressures that severely inhibited the free development and enjoyment of the human body.[25] That it took the insight of the one man who was most interested in recognizing the message of the body to decipher the meaning of these overt symptoms suggests the extent to which the dominant sexual morality of the age suppressed understanding of basic corporeal needs such as free movement and physical pleasure. In a subsequent chapter, I will argue that a central part of Freud's contribution to European culture was his discovery of the meaning of the body, for which the Schreber case supplied abundant material.

CHILD SEXUALITY

The corporeal focal point of parent-child relations in the Victorian period was child sexuality, in particular masturbation. Early masturbatory activity is a function of instinctual impulses, but as parental authority begins to turn upon this early manifestation of child sexuality, it becomes an expression of defiance, an assertion of independence. In a period when sexual morality was particularly severe and when the imposition of parental authority absolute, the playful self-indulgence of masturbating children was especially threatening to parents who were so anxious about suppressing their own sexuality. As soon as the child was able to acquire a moral sense, the parents sought to quash child sexuality with firm moral injunctions, threats of punishment, and, if necessary, physical restraint.

Conflict over the suppression of child sexuality was most acute in the father-son relationship. Perhaps the tyrannical father was partly acting out his anxiety from the loss of his own potency, jealous over the willful assertion of his son's seemingly abundant sexual energy. The combination of fear and jealousy may account for some of the intensity of this struggle in the mid-nineteenth century, but these two factors are universal and not unique to this period. Several other factors may help explain why the effort to control child sexuality was so intense at this time.

The severely restrictive sexual morality of the nineteenth century was imposed by the newly triumphant bourgeoisie. Waste, whether financial or biological, came to be viewed as evil, and the class sexual morality called for even more thrift in the sexual sphere. Child sexuality was a persistent reminder of the human tendency to squander, and parents sought to control it both to instill a proper morality in the child and to reaffirm the wisdom of their own self-restraint.

René Spitz has developed another line of reasoning about the probable causes of the sudden imposition of severe parental authority in curbing masturbation around 1850. From a survey of 559 works relating to the causes, effects, and controls of masturbation throughout history, Spitz concluded that in the late eighteenth century a decisive change of emphasis occurred as physicians began to abandon more moderate treatment involving bland diets and counseling and turned to more sadistic restraining methods. "While in the eighteenth century medical men endeavored to *cure* masturbation, in the nineteenth century they were trying to *suppress* it." [26] In the period 1850–1879 the most sadistic measures involving surgical operations (punitive circumcision, male and female infibulation, and clitoridectomy) were recommended more frequently than other procedures. From 1880 to 1904 various restraining apparatuses were the most frequently used control method, while after 1905 the spread of knowledge of psychoanalysis gradually began to reduce the incidence of surgical and physical suppression of child sexuality. Spitz suggests several historical factors which contributed to the particularly severe measures of the mid and late nineteenth century. The rise of Protestantism placed unprecedented responsibility upon the father to impose moral restraints upon himself and his family. The philosophy of the Enlightenment underlined the need for public health to be a major concern of the state. The connection between the familial responsibility of the father to watch over the morality of his family and the civic responsibility of the state to control public health was realized in the early nineteenth century as the social problems created by the industrial revolution further illustrated the need for societal support of public health projects. By the early nineteenth century, medical men began to accept the social responsibility of medicine.

The historical record confirms Spitz's conclusion that "the sadistic trend in anti-masturbationist therapy came at a period in history when people became aware of infantile sexuality." [27] Parents intruded their anxious moralizing into the most intimate biological and emotional processes of their children in order to curb possible sources of sexual excitation. They were counseled to keep their children to a strict sleeping regimen in order not to allow them to linger in bed either before falling to sleep or after awakening in the morning. Children's diets were carefully controlled to exclude any spicy foods which were believed to arouse erotic impulses. The bowels were carefully watched because constipation and piles were listed among the causes of masturbation. Clothing had to be specially designed. It was suggested that suspenders be loose to prevent pants from rubbing against the genitals, and one moralist advised carefully selecting boys' trousers with shallow and widely separated pockets. Horseback riding, prolonged sitting, and bicycle riding were to be curtailed because they too could generate masturbatory impulses from rhythmic friction. Of course the child's reading must be carefully regulated, and even many suggestive works of art could be dangerous. Other moralists mention as possible exciting causes the easy accessibility of closets, perfumes, furs, or rocking chairs. Thus parents could meddle in every aspect of a child's life under the pretext of rooting out potential causes of masturbatory activity. As if in a final gasp of anxious frustration, the American psychologist G. Stanley Hall wrote that the hand itself, that symbol of man's achievement over the animal world, might prove to be the cause of his ruin.[28] And for the child who continued to masturbate in defiance of the more subtle preventative efforts of his parents, the Victorians resorted to force. Wandering hands were tied down at night, and a variety of elaborate restrictive devices were tried. Lafonde's imposing corset restricted access by a metal cup over the genitals. Some of the extreme measures used to stop a young man from masturbating illustrate how deeply the Victorians feared the ravages of male sexual excess. The medical literature mentioned a variety of procedures such as applying ice on the genitals at bed time, cork cushions placed on the inner side of the thigh to prevent involuntary masturbatory rubbing during sleep, or tying hands to the bed posts.[29] Others were subjected to blood-letting or various kinds of punitive treatment which involved blistering of the penis or removing part of the foreskin as a symbolic punitive castration.

In the late nineteenth century there developed a growing interest in the forces that held the family together. We must assume that this intellectual development reflects historical fact, that these binding forces were deeply felt. At the same time many feared that the institution was falling apart. Partial knowledge of the mechanisms involved in infectious diseases and in hereditary and cultural transmission suggested that the members of a family were inextricably united by invisible chemical substances that flowed from one to another. While the evidence that children inherit their parents' vices and deficiencies abounded, there was far less to indicate that they inherited positive characteristics. Francis Galton concluded from his researches that children inherited a tendency for genius, but far more in keeping with the times was the belief that insanity was inherited. As social scientists observed and explained the process of cultural transmission more exactly, they adduced more evidence that family life determined the temperament and character of children. In responses to the many social, economic, and cultural forces that held the Victorian family together as a sanctuary from the rest of the world, many attempted to break away. The literature reflects this dialectical process: a growing sense of the biological and psychological unity of the family and a mounting tension between the members who sought independence from it.

One quotation well illustrates the kinds of pressures that held the Victorian family together and that ultimately began to tear it apart. It is from a marriage manual by the American phrenologist O. S. Fowler, *Love and Parentage Applied to the Improvement of Offspring*. The passage concludes a section on the nature of spiritual love in marriage and shows to what extent the Victorian parent was led to believe that the family unit was held fast by the innermost biological and psychological processes of man. The italics are Fowler's.

> While sexual love, as such, transmits the bodily organs and animal functions, it remains for this *spiritual* love to call forth into the most delightful and intense action possible, the entire *intellectual and moral* nature of parents, preparatory, and in order, to its conferring on man this boon of *angels,* this "image and likeness" of God; besides purifying and sanctifying the animal by the ascendency of the moral, and guiding all by *reason.* And it is this *combined and concentrated,* as well as *high-wrought,* inter-communion of *every physical, every intellectual, every moral element and function of humanity* in generation as it is *by constitution,* which renders the pleasure attendant on this *double* repast so *indescribably* exalted and beatific to those who *spiritually love* each other, or in proportion thereto; besides being *the* ONLY means of augmenting and perfecting the intellectuality and morality of its product—redoubling more and more as its handmaid love becomes more and more perfect, and thereby enhances, and also unites, in this holy alliance, faculty after faculty, till finally, when both love and generation have their perfect, and of course *united,* work, they embrace within the wide range of their sanctified enjoyment, every animal, every intellectual, every moral, organ and function, and element of man's entire constitution! [30]

The syntax of the passage forms a crescendo of the many themes Fowler had been developing and, laced with italics and prolonged as though the author could not bear to conclude each of its two sentences, strains to convey the enormity and totality of the marital union, culminating in offspring which are a product of *every* physical and mental attribute of the parents. Fowler is relentless in his apotheosis of the sanctity of the family. He implies that life outside it is immoral, irrational, and base. The marital union is the center point of Victorian life, where all is "combined and concentrated," where "every intellectual, every moral element," every "organ and function," is *"united"* totally and for all time in view of the angels, in the "image and likeness" of God. In this world divorce is unthinkable—a rejection of God, an abandonment of the protective hand of religion, a violation of the natural forces of human biology, a corruption of the intellectual and moral qualities of mankind, and a serious threat to the well-being of the state. There is a stickiness here that made family life as unbearable as it was vital to its members who had become so dependent on it. By the end of the century the family had become an explosive unit, poised in a tenuous solidity that was already beginning to decompose as the very forces and tensions that held it together became too much for its individual members to bear. Those tensions were in part a consequence of a growing awareness of the bodily determinants of human existence, in particular as they influenced family life in the late Victorian period.

Chapter 11

The Scientific Study of Sex

The way human sexuality is investigated in France, England, Germany, and America reflects some clichéd characteristics of the intellectual climate of each country. The French have traditionally relied on their great imaginative writers—Rousseau, Stendhal, Balzac, Zola, Proust—to explore male-female relations and have not dwelled on the anatomical details of the sex organs and workings of the sex impulse so thoroughly as many English and German scholars. The English produced a great number of empirical and practical "how to do it" manuals for young men and women. The Germans wrote elaborate theoretical and systematic works classifying and speculating about the various stages in the development of the sex impulse, its component parts, and its significance for human existence. Though the French and the English did make some substantial contributions to the scientific study of sex, the Germans firmly established it as a subject for serious investigation. The term *"Sexualwissenschaft"* (science of sex) was introduced by the German researcher Iwan Bloch in 1906. The contributions of Charles Féré and Alfred Binet in France and even Havelock Ellis in England pale next to the prodigious works of Richard von Krafft-Ebing, Albert Moll, Iwan Bloch, Magnus Hirschfeld, Wilhelm Bölsche, Otto Weininger, Wilhelm Fliess, and Sigmund Freud in Germany and Austria. The Americans developed the statistical approach to the study of sex, and I shall conclude this chapter with an account of the work of the man whose name is synonymous with this kind of study—Alfred Kinsey.

Studies of sex appeared regularly throughout the nineteenth century, but beginning with Krafft-Ebing's *Psychopathia Sexualis* in 1886 the systematic study of sex proceeded at a quickened pace. A number of sub-topics were pursued by independent investigators all over Europe. There were three general areas of investigation: the biology of the sex organs and the reproductive process, historical and anthropological studies of different sexual customs, and clinical studies of the sex impulse and its influence on normal and abnormal mental development.

In the last thirty years of the nineteenth century European scientists came to understand a great deal about the nature of sexual reproduction. In 1875 a Berlin physiologist, Otto Hertwig, discovered the essentials of the process of fertilization—the union of male and female sex cells. In 1879 Hermann Fol made the first microscopic observation of a sperm cell penetrating an egg.[1] In 1887 the Belgian Edouard van Beneden discovered that in the formation of an egg the pairs of chromosomes split and thus produced sex cells with half the number of chromosomes of other cells; and in the same year August Weismann discovered that prior to their union the male and female sex cells divided in two, thus enabling chromosomes from the parent cells each to contribute to the heredity of the offspring. Though the details of these biological processes were not of great interest to the larger public, some general facts about human sexuality gradually crept into the popular studies. Following up on the research of Weismann in particular, the Lamarckian notion that acquired characteristics were inherited began to give way after 1890 to the idea that the sex cells were independent of the body cells and that hereditary endowment operates independently of the life experience of parents. Thus parents' virtues and vices were *not* inevitably passed on to children.

The complicated nature of sexual reproduction was simplified and popularized in a number of studies, the most elaborate of which was Wilhelm Bölsche's interminable rhapsody on the beauty of sexuality, *Love-Life in Nature* (1898, 1902). In these two large volumes of lyrical biology Bölsche recreated the entire evolutionary process derived from a fundamental sexual impulse, beginning with asexual reproduction of unicellular organisms and culminating in the poetry of Goethe. In the beginning "the strength of a God in a high hour of consecration once poured out infinite semen of all living things into the dead barren waste of the ocean." The essential and irrepressible impulse of sexuality drove life on to higher forms and increasingly refined transformations of this initial impulse. "The whole colossal savage substructure of sexual love from the beast upward, from the fish and the day fly, was necessary for the organic evolution of man's great creation, human love." [2] And so "blood became spirit." Nietzsche's remark that "there is more reason in your body than in your highest wisdom" offered a point of departure for Bölsche's theory that while the individual mind may be but a few years old, each bodily cell has a memory that goes back aeons. We ought, therefore, to learn from our bodies this ancient lore which is locked in every cell. Bölsche's eulogy on the beauty and goodness of all creation even included the human excremental function, and he railed against contemporary prejudices about the anus and excrement. Why, he asked, should we regard the anus and excrement as associated with hell, the devil, and evil? Rather, we should "allow ourselves to learn from the anus and its wisdom *[vom After und seiner Weisheit]*, as from that of the mouth." Without the anus, man would have to get rid of waste material through the mouth like lower animals, and, he punned, we should therefore learn to appreciate *Afterweisheit* (sophistry—but literally, the wisdom of the anus).[3]

Bölsche presented to the general reading public the latest discoveries in the biology of sexual reproduction transformed into stories choked with symbol and metaphor. He depicted the fertilization of an egg as he imagined it appeared from within a woman's body. "An elastic ball . . . glides slowly along a field of wheat. . . . Minute particles seem to be swimming about in it. . . . We think we can count twenty-four strands." Strange, pear-shaped little monsters approach the great ball and throng about it, and after one penetrates it "those who remain outside of the new sheath must, often waiting hopelessly in the ante-chamber, finally perish like wretches." [4] In such a manner the German public learned that the female egg was moved through the Fallopian tubes by cilia, that there were twenty-four chromosomes in each sex cell, and that a protective sheath was created about a fertilized egg to prevent the penetration of any other sperm cell. The interior anatomical landscape of the human sexual apparatus thus became the setting for romantic biology. However strained Bölsche's literary devices appear today, they offered a highly repressed population an opportunity to dwell upon those bodily processes which had for generations been forbidden topics for conversation or even for thought.

Another area of sexual biology that received widespread popular interest was the study of hormones, in particular their influence on sexual potency. In the 1850s Claude Bernard had begun to study the effect of "internal secretions" on bodily processes, but a great deal of interest was generated in the subject when his successor at the University of Paris, Charles Brown-Séquard, reported in 1889 that injections of hormones from frogs and dogs had a rejuvenating effect on man. The German biologist Eugen Steinach developed this process and claimed to have discovered the part of the testicles which controls the "manly function." He devised an operation which involved transplanting extracts from that part of animal testicles to produce sexual rejuvenation in men. Steinach claimed to have prolonged the normal life span of laboratory rats with his "rejuvenation" operation, and he artificially created hermaphrodites with hormone injections. His hormonal therapy reached a high point of popularity in the 1920s, but the inconclusiveness of the results gradually led to a loss of interest.

As the scientific study of sex gained momentum around the turn of the century, the informed public found it increasingly difficult to discern genuine medical advance from charlatanry and quackery. And if Steinach's injections of "monkey glands" did not restore potency as promised, at least they did not do the kind of damage that mercury injections produced in the treatment of syphilis. The promise of a sexual fountain of youth must have been most alluring at a time when the demands on male sexual potency were being made by women whose own sexual needs had been suppressed throughout the century.

Biology and psychology are combined in research on the sexual impulse and its excitation in various bodily zones. Freud presented his theory of erotogenic zones in his full developmental psychology of 1905, but the medical literature of the late nineteenth century offered a wide range of striking anticipations of his thought on the components of the sexual impulse and on erotogenic zones.

An early analysis of the sexual impulse into components was made by Herbert Spencer in 1855. He listed nine elements of it which included the physical impulse, the proprietary feeling, the feeling for beauty, and self-esteem. He concluded that the impulse ultimately "fuses into one immense aggregate of the elementary excitations of which we are capable." [5] In 1894 a German professor of gynecology, Alfred Hegar, distinguished between two components of the sex instinct—the urge to copulate *(Begattungstrieb)* and the urge to reproduce *(Fortpflanzungstrieb)*. This distinction was accepted and elaborated upon in the following year by Alfred Eulenberg in *Sexuelle Neuropathologie*. Also in 1894 Max Dessoir described the "undifferentiated sexual feeling" as a component that appears just prior to puberty when sexual feelings are not concentrated in the genitals and seek only "contact with a warm body." This component of the sexual impulse may in normal individuals last for a few years beyond puberty but gradually gives way to genital, heterosexual feelings.[6]

In 1898 a Berlin physician, Albert Moll, described two components of the sexual instinct: the *Detumescenztrieb*, an impulse to discharge sexual excitations in the genitals, and the *Kontrektationstrieb*, an impulse to contact another person. Moll analyzed a variety of sexual activities as functions of these component parts. In homosexuals, for example, the *Detumescenztrieb* is stronger than the *Kontrektationstrieb*, while in women who suppress the desire for genital stimulation the absence of the former drive produces sexual anesthesia.[7]

In 1903, in a popularization of medical theories for the layman, Hans Rau noted that "sexual enjoyment is made up of a series of emotional components," but the only specific component he mentioned was the feeling of pain. This connection between pain and sexual feeling, he believed, explained perversion: one of the components becomes excessively intense, dominates the others, and produces a confusion of instincts. This inclination to perversion, he believed, slumbers in everyone.[8] In 1904 a German physician, Friedrich Scholz, wrote that the sense of smell and the desire to expose oneself were "essential parts of sexual enjoyment." [9] This theory, like that of Rau, Krafft-Ebing, and others, helped expand the meaning of sexuality beyond the limited conception that it was solely rooted in the genital function. The idea that certain bodily zones are particularly well constructed to generate sexual excitation is generally associated with Freud's theory of erotogenic zones—oral, anal, and phallic—but this idea was a contribution of a number of scientists of the late nineteenth century and forms another part of the vast accumulation of information about human sensuousness at that time.

The history of the term "erotogenic zone" begins with Jean Charcot's *Leçons sur les maladies du système nerveux* (1872), where the term *"zones hystérogènes"* was used to describe particular bodily regions which, when stimulated, precipitated conversion forms of hysteria (fig. 22). The cutaneous, mucous, and visceral zones were designated as having this property. In 1881 the French psychiatrist Ernest Chambard introduced the term *"centres érogènes,"* which he compared to Charcot's hysterogenic zones in that when stimulated they generated sexual excitation that could produce orgasms. The particular bodily zones Chambard included in this category were the genitals, the inner side of the thigh, the groin, the nipples, the neck, and the palms.[10]

The first use of the precise term *zones érogènes* was made in 1883 by the French physician P. Kerval in a review of the work of Féré, Charcot, Richter, and Chambard on hypnotism. Referring explicitly to the term used by Chambard, Kerval wrote that "certain hysterics have bodily regions, *zones érogènes,* which are not without analogy with *zones hystérogènes* and which provoke genital sensations intense enough to produce an orgasm when they are stimulated during a somnambulistic state." [11]

In 1885 Albert Pitres reported that these zones could induce hypnotism as well as hysterical states.[12] The term was introduced to a wider public by Alfred Binet and Charles Féré in *Animal Magnetism* (1887). After describing the function of the hysterogenic zones in accord with Charcot's original definition, they explained that "in the case of some hysterical subjects there are regions in certain parts of the body termed by Chambard *zones érogènes* which have some analogy with the hysterogenous zones, and simple contact with these, when the subject is in a state of somnambulism, produces genital sensations of such intensity as to cause an orgasm."

In their enthusiasm they said, without evidence, that the erotogenic zone might be transferred by a magnet and must be stimulated by a person of the opposite sex to be effective. In a popular manual, *The Sexual Instinct* (1899), Féré again explained the function of the erotogenic zones, this time adding that they could be effective in a normal state, not just in somnambulists, hysterics, and hypnotized patients.

The term entered German literature through the works of Albert Moll and Krafft-Ebing. Moll described the function of "the so-called *zones érogènes*" in which "stimulation of certain areas of skin causes pleasurable sensations that are reflected in the sex organs." [13] Iwan Bloch explained that a passive experience of pederasty in childhood could cause an abnormal condition in the anal region that might become an erogenous zone. And, he added, "other contacts with the anal region, especially flagellation, can create an erogenous zone, which may lead to homosexuality in subsequent sex life." [14] By the time of the publication of Freud's *Three Essays* in 1905, the idea of erotogenic zones was discussed by numerous European scholars, who speculated that the anus, the sex organs, and other areas of the skin had special sensitivity to sexual stimuli.

In addition to studies of the physiology of sex, a number of popular works began to compare European sexual customs with those of other cultures and other periods. These anthropological and historical works tended to question the value and quality of European sex life at a time when moralists and social critics were also beginning to explore "the sexual question." The German anthropologist Heinrich Ploss collected material on women in many primitive cultures for his monumental *Das Weib* (1885). Ploss's research inspired subsequent studies in comparative sex ethics such as Mueller-Lyer's multi-volume work on love and marriage. In 1908 the German geographer Otto Stoll published a massive anthropological study of sex life in folklore.[15] Freud's first full-length psychoanalytical history of sexual customs, *Totem and Taboo* (1912–1913), derived from this new and flourishing area of German scholarship. The two great English studies in this genre, Westermarck's *History of Human Marriage* and Havelock Ellis's *Studies in the Psychology of Sex*, were also intended to help evaluate current European sexual and marital customs and expand the prevailing constraints on sexual relations.

Havelock Ellis compiled seven volumes of material on topics ranging from child sexuality to adult sexual pathology. His own bizarre sexual development no doubt contributed to his fascination with normal and abnormal sexual life. At the age of twelve his mother, thinking Havelock was not looking, paused to urinate while standing on a remote path of the London Zoological Gardens. Thereafter, he reported in his autobiography, he had "a slight strain" of what he called "urolagnia"—sexual excitation from observing a woman urinating, preferably in a standing position. At seventeen he began to keep a record of his nocturnal emissions, because he believed they were a sign of "spermatorrhea." The source of his thinking about this disease was the work of Charles Drysdale, who had argued that the progress of the disease eventually debilitated the entire male sexual apparatus.[16] During his medical studies in the 1890s he had a romance with the novelist Olive Schreiner, who much enjoyed observing living sperm under a microscope. Ellis supplied her with both the sperm and the microscope, although when he married Edith Lees in 1891 he was still a virgin. The next year he discovered that his wife was having a homosexual affair, and he himself began to have his own heterosexual affairs—one of which was with Margaret Sanger, a leader of the birth control movement in America, and another with Françoise Delisle, who has left a vivid account of his urolagnia and his occasional impotence.[17] In 1898 his wife published *Kit's Woman*, a novel about a love between a woman and an impotent man which was partly autobiographical and which publicly announced Havelock's inadequacies.

Ellis thus experienced a number of the anxieties and insecurities that plagued European men in the Victorian era. His acceptance of Drysdale's theory about the dangerous consequences of pollutions led to his self-imposed cessation of masturbatory sexual release. His urolagnia was symptomatic of the many sexual fetishes and perversities that flourished in the Victorian period, and his marital failure—a mixture of rejection, homosexuality, and impotence—completed the tragic scenario. The motive behind his massive, almost compulsive, cataloging of sexual practices was the need to understand and explain his own struggle to achieve sexual gratification. It is not surprising that the general message of the collection is a passionate critique of sexual repression and a demand for greater sexual freedom. Ellis enthusiastically endorsed the campaigns for sex education, clothing reform, nudism, contraception, freedom of divorce, toleration of homosexuality, and the liberalization of censorship laws.

His research drew from a host of medical and popular studies of sex that had appeared in the late nineteenth century. A perusal of his sources shows that German, French, and English scholars had produced a vast store of biological, anthropological, and clinical information about every aspect of human sexuality. For the most part these works covered very similar material, differing primarily in the moral judgment each adopted. Three issues were the most hotly debated: masturbation, adult sexuality, and homosexuality. Some saw masturbation as the cause of countless physical and mental diseases; thus Hermann Rohleder, whose compendium on the causes, effects, and treatments for masturbation constituted a 320-page treatise.[18] Even Freud in his early clinical studies regarded masturbation as pathogenic. But many physicians began to regard moderate masturbation as a healthy outlet for sexual frustration, and one even came to view it as "a training school for the future." [19]

Against the long-held theory that sexual intercourse debilitated men, some writers reported on the benefits deriving from a regular discharge of sexual substances. Albert Moll took this position in his study of 1898, which was intended to provide a corrective to the preoccupation with sexual pathology which, he complained, so dominated the most recent studies of the sexual impulse.

The third most controversial issue—homosexuality—attracted the interest particularly of German scholars whose defense of homosexuality was focused on achieving repeal of the notorious Paragraph 175 of the German Penal Code, which provided severe punishment for male homosexuality. The studies of Iwan Bloch and Magnus Hirschfeld were devoted to creating a more tolerant attitude toward homosexuality in German society and to assisting homosexuals to live more comfortably with their own sexual predilections. The work of Iwan Bloch in Germany most closely paralleled that of Havelock Ellis in England, both in its scope and in its moral tone. This Berlin physician was first drawn to the study of sexual pathology and perversion. His *The Marquis de Sade and His*

Time (1899) examined the historical origins of sexual sadism, and another work of 1902 was a clinical study of sexual pathology. In 1906 he assembled a mass of research for his comprehensive study *The Sexual Life of Our Time*, in which he coined the term "science of sex" and inaugurated an era of serious study of human sexuality. He explained that this work was the product of a ten-year effort to create a science of sex, and a community of eager and courageous scholars followed his lead. Bloch attempted to get beyond the "purely medical" conception of sexual life that prevailed and use the methods and materials of anthropology and cultural history. He attempted to integrate social and economic factors in his study to bring it in line with the views of the newly founded periodical *Mutterschutz* (1905) whose interests were sex education, contraception, reform of conventional marriage, and greater tolerance for prostitutes. In 1912 Bloch produced a lengthy study of prostitution and in 1914 founded the *Zeitschrift für Sexualwissenschaft*, which provided a forum for the new science of sex until it was banned in 1932. His work also helped provide the theoretical foundation for the *Institut für Sexualwissenschaft*, which was founded by Magnus Hirschfeld in 1919. It too fell prey to a vicious attack by the Nazis. On May 10, 1933, during the burning of the books in Berlin, the Institute was destroyed and the head of Hirschfeld was carried in effigy.

In the early decades of the twentieth century Germany led in the effort to establish a systematic and objective study of human sexuality. Though none aside from Hirschfeld and Freud achieved the status of Bloch, a number of other important contributions made a great stir in their time and added to the steadily growing understanding and enjoyment of sex. In 1902 the twenty-two-year-old academic prodigy Otto Weininger received his doctorate in philosophy at the University of Vienna for his thesis *Sex and Character*. The day his thesis was accepted he converted from Judaism to Christianity, and in 1903, just after his thesis was published, he committed suicide. Wilhelm Fliess accused him of plagiarizing his theories of bisexuality and sexual periodicity, and Paul Möbius claimed that Weininger had stolen his theory of the physiological feeblemindedness of women. Weininger's early publication, his dramatic religious conversion, his suicide, and his posthumous involvement in accusations of plagiarism gave the book enormous publicity and helped make it a lasting best seller.

Weininger began by arguing that the sexual impulse is not limited to the sex organs but is distributed throughout the entire body. He then introduced the theory of bisexuality, according to which everyone has remnants of an original bisexual disposition. Before five weeks the sex of an embryo cannot be determined, and thereafter sex differentiation is never entirely complete. According to his "Law of Sexual Attraction" a woman with the sexual composition of one-quarter male and three-quarter female will most likely be attracted to a one-quarter female and three-quarter male. This kind of thinking led to his critique of the view of Krafft-Ebing that all instances of sexual inversion are pathological. Weininger insisted rather that the traces of undifferentiated sexuality in both sexes may become manifest in normal homosexual impulses.

After vehemently defending his theory of normal human bisexuality, Weininger rehashed a number of current theories that maintained the sharp opposition of the sexes with women generally being dubbed inferior. For the man, puberty is unexpected and disgusting, while for the woman it is a long-awaited fulfillment. Women can experience sexual stimulation anywhere on their bodies and at any time, whereas man's sexual capacity is limited to specific intervals and is localized in the genitals. These sexual differences are reflected in personality. In a section no doubt lifted from Möbius's vicious misogynist tract *The Physiological Feeblemindedness of Women* (1901), Weininger wrote: "The greater articulation of the mental data in man is reflected in the more marked character of his body and force, as compared with the roundness and vagueness of women." [20] The connection between body and character is further exemplified by woman's higher disposition to hysteria, which he explained as "the organic crisis of the organic untruthfulness of women," that is, of their attempt to adopt men's values about sex.[21] He then took up the problem that obsessed August Strindberg—that no man can know for certain if he is the real father of his child. Women use this secret and intensify the dishonesty inherent in their sex. Weininger believed that the reproductive function of women further tied their personality to anatomy. Woman is "impregnated not only through the genital tract, but through every fiber of her being. All her life makes an impression on her and throws its image on her child."

Following his insistence on the psycho-organic passivity and amorphousness of woman, Weininger began to argue for her servility. "Woman is essentially a phallus worshipper, . . . permeated with a fear like that of a bird for a snake." He elaborated upon this idea with a striking anticipation of Freud's theory of penis envy: "It has never until now been made clear where the bondage of women lies; it is in the sovereign, all too welcome power wielded on them by the Phallus." [22] In a final section Weininger speculated on what he believed to be the unique nature of Jewish sexuality. As we shall see in a later chapter, the fear that Jews had special sexual powers would haunt the Nazis. Weininger supplied some early ingredients for this fear by arguing that the Jew is "always more absorbed by sexual matters than the Aryan," although he was quick to add that the Jew was "notably less potent sexually and less liable to be enmeshed in great passion." [23] He concluded that Jews, like women, had a slave mentality—both lacked a soul.

Fliess and Möbius openly accused Weininger of plagiarism, and there can be no doubt that this young scholar drew heavily from others. His debt to Havelock Ellis was enormous. But his book nevertheless shows much originality, particularly in working out possible channels of influence between the biology of human sexuality and character. His specific conclusions were laden with the antifeminist prejudices of male scholars of his age, so threatened by the onslaught of emergent female sexuality. But his style of thinking expanded discussion about the influence of bodily structures and functions on human existence precisely when Freud was working out his own theories of the anatomical basis of personality. The confluence of these studies shows that the desire to work out, as Weininger had called it, "a biology of ideas" was shared by many and constitutes another element of the growing recognition of the importance of the human body.

August Forel's *The Sexual Question* (1905) was the most popular book on sex to appear anywhere in Europe before the First World War. By 1920 it was reprinted in a thirteenth edition, and it was translated into French, English, and Italian. It was largely a summary of the recent clinical and theoretical work on the study of sex and argued for a "clean sweep of prejudices, traditions, and prudery." This respectable Swiss psychiatrist gave his readers the liberal moralizing that they sought. The introduction set the affirmative tone: "sexual life is beautiful as well as good." After surveying the by-then standard subtopics—the biology of sex, sexual pathology, venereal disease, prostitution—Forel proposed two major goals for sexual reform. The negative task was to suppress the causes of "sexual evil." He argued for the elimination of the "cult of money" which governed certain unfortunate marital decisions and sustained prostitution, suppression of women, profits in pornography, and venereal disease and which further threatened the Aryan race with "extermination by the fecundity of other races." [24] The positive task was racial betterment. Certain undesirable types ought to be eliminated from the racial stock: criminals, lunatics, imbeciles, alcoholics, tuberculars, and even "all who are irresponsible, mischievous, quarrelsome or amoral." [25] This mixture of progressive physical culture and racial chauvinism anticipated a good deal of the literature that became influential during the Nazi period.

Before turning to the study of sexual pathology and homosexuality, which was the complement to these studies of normal sexuality, I would like to comment on the work of Alfred Kinsey and the historical significance of the quantitative approach to the study of sex.

Sexual Behavior in the Human Male (1948) was an instant best-seller. At last a man could find out the "truth" about sexuality based on quantifiable data. He could learn, for example, that the average husband had experienced 1,523 orgasms before marriage and the average wife only 223. The book was useful for determining class differences in the length and kind of foreplay, the frequency of marital impotence, or pre-marital sexual activity. The middle and upper classes, Kinsey demonstrated, engaged in more elaborate sexual experiments than the lower classes. Against the traditional idea that the lower classes have the fuller sex life, he argued that middle-class women have a higher likelihood of attaining orgasm than working-class women. Also surprising was the seemingly high incidence of homosexuality—6.3% of the population.

The book reveals the total abdication of all considerations of quality in sex life. The pervading ethic is that orgasm frequency—no matter how, with whom, or with what—is an absolute good, a measure of sexual success. For the curious, Kinsey offered a breakdown of the sources of orgasms among the entire population. He informed his readers that 6.3% of all orgasms come from homosexual contacts, 69.4% from heterosexual contacts, 24% from masturbation or nocturnal emissions, and 0.3% from contact with animals. These figures were based on the say-so of his by-now immortal 1,200 interviewees.

Unlike the Victorian man who might have been tormented by a vague sense of his sexual inadequacy, the reader who accepted Kinsey's figures could precisely measure the extent to which his sexual performance fell short of the norm. And those men who discovered that the number of their "sexual contacts" exceeded the normal weekly average might well have cut back to spare their partners unwanted sexual excess. The concept of "a normal sex life" at last came to have some exact meaning determined by the statistical averages of Kinsey's report. Beginning in 1948 sexual performance acquired a measurable value unprecedented in the history of human relations. The era of the scientific study of sex had culminated in a new era of the scientific experience of sex. Human sexuality became a competitive event subject to the same kinds of evaluations on the basis of speed, frequency, and endurance previously reserved for athletic events and machine production.

Chapter 12

Sexual Pathology and Homosexuality

Human knowledge sometimes progresses on its negative side. Over the centuries Christian thinkers have never tired of reminding mankind of the potential sinfulness of sex, and philosophers have entreated their readers to consider the immorality of it, but in the late nineteenth century a new emphasis emerged—its danger to physical and mental health. The title of the classic of this genre—Freiherr Richard von Krafft-Ebing's *Psychopathia Sexualis* (1886)—drew the connection forthrightly. It was a study limited to sexual pathology, but the overwhelming impression the book gave from the title to the conclusion was that sex was more often linked with disease than with normal functions. The subtitle, "With Especial Reference to the Antipathic Sexual Instinct, a Medico-Forensic Study," made explicit the pathogenic potential of sex. Although his focus was on the diseases of the sexual impulse and although his interpretation was alarmist, his work was symptomatic of the growing interest in the human body in general and the willingness to devote scholarly attention to it.

Krafft-Ebing began by claiming that "few people are conscious of the deep influence exerted by sexual life upon the sentiment, thought and action of man in his social relations to others." He cited Schiller's famous remark, "While philosophers dispute, hunger and love decide our fate," and added with some daring, "Sexual feeling is really the root of all ethics, and no doubt of aestheticism and religion." [1] But then he considered the consequences of animal-like lust which destroys all "honour, substance and health." Christianity has helped tame this instinct by exalting the virtues of chastity, virginity, modesty, and sexual fidelity; but the exaggerated stress placed upon the nervous system by modern urban life is threatening to undo the accomplishments of Christian asceticism and allow an eruption of long-suppressed animal instincts. The reduction of masculine virility may lead to further sexual pathology. Love can only exist in a healthy state when it occurs "between persons of different sex capable of sexual intercourse." Therefore, the burden of responsibility for maintaining healthy sexual relations falls upon men, because their sexual impulse is stronger and because normal women ought to remain sexually passive. Krafft-Ebing then revived that opposite view of female sexuality that we have observed in a number of writers of this period. "Nevertheless, sexual consciousness is stronger in woman than in man," because her need of love is greater than man's. Female sexuality is further trivialized by his theory that the wife accepts marital intercourse largely for "proof of her husband's affection." [2] If Krafft-Ebing was correct about this aspect of female sexuality, then men must have been subjected to enormous pressure to perform adequately. They not only had to strain to please their passive wives (a difficult, if not impossible, task), but their failure to perform was evidence of infidelity. Another burden on men was that the intensity of their sexual desire made them dependent upon women. This sexual dependency is the mainspring for the masochistic sexual relations described elsewhere in the text. He viewed male sex life as largely a source of anxiety—a test of masculine power, a test of fidelity, and a source of enslavement to women, who themselves were not driven by an eternally demanding sexual impulse.

Krafft-Ebing then considered a certain kind of sexual pathology which would occupy him through a good portion of the book—fetishism. He defined a fetish as an object or part of a body which by means of association could arouse sexual excitation and produce orgasm. The hair, hands, or feet of women commonly exercised fetishistic powers over men and became the exclusive object of their sexual interests. He explored both hereditary and environmental causes of fetishism. He argued that "pathological fetishism seems to arise only on the basis of a psychopathic constitution that is for the most part hereditary,"[3] but he was more interested in the environmental causes which had been explored by French scholars. In 1882 Charcot and Magnan had explained sexual abnormalities as a consequence of traumatic experiences from early childhood. One famous example they cited concerned a young man who could not successfully have intercourse with his wife unless she donned a nightcap similar to one he had seen his mother wear when as a boy he used to get in bed with her.[4] Alfred Binet developed the thesis that sexual fetishes often focus on articles of women's clothing, since such fetishes have their traumatic origins in childhood—a time when the opportunity for normal intercourse is not yet available, and when early stirrings of the sexual impulse find only clothed women as their object.[5]

The theory of fetishism developed by Binet and Krafft-Ebing reveals a significant extension of the conception of sexuality beyond exclusively genital activity. Freud would soon extend that range even farther and argue that the energy driving the entire psychic apparatus is sexual, but these early speculations reveal a new and broader interpretation of the erotic potential of the human body.

Krafft-Ebing devoted a large section to fetishes involving smell.[6] Though other psychologists of smell argued that olfactory sensations were normal, indeed that they constituted an important ingredient in sex life, Kraff-Ebing maintained that sexual excitation produced by olfaction must be viewed as pathological. His critical evaluation of the function of smell in sex is, however, of little historical consequence—what is important is that he unwittingly offered a new range of sensory experiences which would be incorporated into the general realm of human sexuality. The various arguments for the reintroduction of the sense of smell in human sexual life offered by Iwan Bloch and Havelock Ellis, and by implication Freud, were the direct historical consequence of Krafft-Ebing's exploration of this aspect of human sexuality. To support this area of his investigation he offered two arguments for the anatomical connection between the olfactory system and sexual excitation. He surveyed literature that demonstrated the similarity between the erectile tissue in the nose, nipples, and genitals, all of which swell in response to sexual excitation. The similarity between the erectile tissue of the nose and female sexual organs was noted by Wilhelm Fliess, and Krafft-Ebing cited this important work in later editions.[7] He further discussed literature on the embryological connection between the development of nasal and genital tissue. By demonstrating the anatomical link between the nose and the sex organs he speculated about the particular paths by which sexuality permeated a variety of zones of the body which were not traditionally regarded as sexual.

After widening the range of sexuality anatomically in connection with the specific function of the olfactory system, he considered a number of other potentially erotic bodily zones which he called "hyperaesthetic zones." He regarded the sexualization of these zones as pathological, but again the important cultural consequence was the suggestion that the sexual impulse was further modifiable and extendable and could ultimately encompass any part of the body. The idea that anal stimulation might be sexual was given elaboration in Krafft-Ebing's most disapproving account of it. Twenty years later, Freud only had to revise Krafft-Ebing's evaluation of this process to derive a large section of his own theory of the sexual significance of the anal region.

After extending the location of sexual excitation spatially through his theory of hyperaesthetic zones, Krafft-Ebing sought to make another extension in time. Sexuality does not always emerge at puberty, he argued, but may be present in early childhood. He viewed this "premature" awakening as unquestionably pathological. One unfortunate girl "practiced lewdness with boys, . . . seduced her four-year-old sister into masturbation, and at the age of ten was given up to the practice of the most revolting vices. Even a white-hot iron applied to the clitoris had no effect in overcoming the practice, and she masturbated with the cassock of a priest while he was exhorting her to reformation."[8] Such case histories fascinated readers who no doubt vividly remembered the bizarre sexual activities described and forgot Krafft-Ebing's tedious condemnations of them.

Another source of the popularity of this study was the graphic reports on sex crimes. Krafft-Ebing described a number of such crimes, including those of "Jack the Ripper," who terrified London society from 1887 to 1889 by cutting the throats of women and then mutilating their bodies. The Ripper frequently ripped open the abdomen and tore out the intestines. In some instances he cut off the genitals and either tore them to pieces or carried them away. Another sex criminal told of his state of mind during the commission of his crimes. "It never occurred to me to touch or look at the genitals or such things. It satisfied me to seize the women by the neck and suck their blood. To this day I am ignorant of how a woman is formed. During the strangling and after it, I pressed myself on the entire body without thinking of one part more than another." [9] His confession of ignorance about how a woman is formed was symptomatic of a culture that was frantically devoted to concealing the female body under clothing, exalting the nature of its contents, and enshrouding it with mystical and mysterious qualities. In an age of faulty or non-existent sex education, it is conceivable that these crimes would occur in greater number and might generate particularly frenzied attacks on the female body such as the ones described in *Psychopathia Sexualis*.

In addition to extending the spatial and temporal limits of sexuality, Krafft-Ebing also explored a variety of qualitative extensions, in particular its association with pain. We have observed that nineteenth-century aesthetics came to recognize the role of the "ugly" in art; an analogous extension was made in viewing sexual life to include pain. In the late eighteenth century the Marquis de Sade had made a case for sexual pleasure being heightened by inflicting pain, and in the 1870s an Austrian novelist, Leopold von Sacher-Masoch, explored the eroticism generated by being tortured.

Sacher-Masoch's classic *Venus in Furs* (1870) tells of the progressive degradation of the masochist Severin at the hands of his torturess, Wanda. Severin argues that the male-female relation is inherently one of potential conflict where "there is only one alternative: to be the hammer or the anvil." [10] He confesses that he sought to make an anvil of himself, and encouraged his wife to make him suffer. Their sadomasochistic relationship finally breaks off with the ultimate humiliation, his being whipped by her new lover while she looks on amused. This novel immortalized the image of the cruel and beautiful female torturer draped in furs, laughing as she brandishes her whip. In *Psychopathia Sexualis* Krafft-Ebing introduced the term "masochism" to describe this particular sexual perversion, which was integrated into a vast nomenclature of sexual anomalies. He elaborated on the theme of degradation to provide an explanation for the current vogue of masochism. It was an outcome of the desire to reverse the traditional roles of the subordinate woman and submit oneself to subjugation by them. The masochist was symbolically atoning for men's abuse of women by reversing the sex roles. The popularity of shoe fetishism mixed with masochism was the desire to be stepped on by a woman. Krafft-Ebing interpreted other perversions such as being whipped, licking the feet of upper-class women, smelling sweat, and even cunnilingus as stemming from this desire for humiliation.

He repeatedly insisted that these perversions only take hold in the hereditary degenerate, but he did investigate environmental causes with insight. The idea that a man or a woman could get sexual pleasure from reversing traditional sexual and class roles is strikingly modern and illustrates an important transitional stage in medical thinking about sexuality from the narrow limits of adult genital sexuality to a broader range of activities.

A number of historians have argued along similar lines that sadism and masochism have specific historical causes which were particularly influential in the late nineteenth century. Conrad Haemmerling's Marxist study (1929) argued that capitalism encouraged merciless competition and was "a school for sadism," because the capitalist's joy is always linked with the defeat of others. A second factor Haemmerling noted was the reactive conflict generated by the aggressiveness of feminist movements. A unique *"Liebeshass"* grew in this time of male effeminization and female masculinization.[11] Several scholars attributed the high incidence of flagellation in England at this time to the practice of flogging in the British armed forces and public whipping in boys' schools.[12]

I believe that we can explain the popularity of *Psychopathia Sexualis* largely from its exploration of the vast potential of human sexual fantasy life, which until this time was restricted to adult genital sexuality. *Psychopathia Sexualis* was a voyeur's holiday for readers eager for vicarious sexual adventures beyond the frontiers of "good taste" and morality. They could indulge repressed sexual fantasies and at the same time gorge themselves on moral condemnation of sexual deviation. Secure behind the screen of their shared disapproval they observed those "others'" immoral and criminal behavior. Much to their surprise, however, one group began to break through that obscurity. "The love that dared not speak its name" began to speak out for the first time in modern history and emerge from the closets of the European conscience, partly because they were forced into the open and partly because they began to realize that it was high time they did so. Krafft-Ebing devoted the largest part of *Psychopathia Sexualis* to the "sexual inverts," whom he viewed as unquestionably pathological. In 1895 the well-known dramatist Oscar Wilde was exposed as a homosexual and forced to confront a hostile public well schooled on Krafft-Ebing's "medico-forensic" condemnations. Wilde lost his personal battle, but his martyrdom helped others to understand the many ways social prejudice caused homosexuals so much misery.

On February 18, 1895, the Marquis of Queensberry left a card at a London club which accused Oscar Wilde of "posing as a sodomite." The Marquis resorted to this public insult to break up the relationship between Wilde and his son, Alfred Douglas. Wilde decided to take legal action and brought charges against the Marquis for criminal libel. In the ensuing trial the defense produced some letters which Wilde had written to Douglas, one of which contained the damaging words: ". . . it is a marvel that those red rose-leaf lips of yours should have been made no less for the music of song than for the madness of kisses." [13] The Marquis further defended his accusation by introducing several homosexuals who testified to having had relations with Wilde. The evidence produced was so damaging that following the Marquis of Queensberry's acquittal, Wilde was arrested and charged with having committed "indecent acts." A second trial began, this time with Wilde accused of having violated the Criminal Amendment Act of 1885, which made homosexual activity in private between consenting adults punishable by up to two years' imprisonment with hard labor. The second trial was inconclusive, but the state pursued Wilde in a third trial which ended with a conviction and the imposition of a maximum sentence, which Wilde served out to the day.

When he began proceedings against the Marquis in 1895 he was a controversial and central figure in English letters, with two successful plays currently running on the London stage; when he emerged from prison in 1897 he was a social outcast and emotionally broken—estranged from his wife and children, bankrupt, and unable to resume writing with his former genius. Isolation and scandal had burned out the cheerful wit and creative energy of a leading cultural figure. His personal tragedy had repercussions throughout English cultural circles. *The Yellow Book* folded, the publication of an important study of homosexuality was impeded, and homosexuals streamed over to the continent to avoid similar harassment.[14] But although the first reactions to the trials of Oscar Wilde were caution and silence, the entire episode left a more lasting source of both outrage and inspiration which others would soon convert into an active campaign for greater understanding and toleration of homosexuality.[15]

The barrage of gossip that followed the revelations of the trials saw new speculations about human sexuality that was at last beginning to emerge from the restrictive bonds in which it had been held throughout the nineteenth century. Fantasies about what Wilde might have done with his own body and the bodies of other homosexuals made an important contribution to the larger view of sexuality, an expansion which, I have argued, began to take place with the growing interest in sexual pathology generated by Krafft-Ebing. The message that remained with those who disapproved of Wilde's activities, as much as with those who argued for tolerance, was that the human body craves a variety of sexual stimuli that often transcends the minimal requirements necessary to achieve sexual reproduction. The scandal was the implication that the body was to be played with according to individual preference. Why should it trouble anyone that Oscar Wilde had enjoyed reaching through special slits in the pockets of young men in order to masturbate them?[16] The prevailing thought was that the body was supposed to be used for very limited sexual activities; but if a leader of the London cultural establishment had broken the rules, it seemed that the codes which regulated human sexuality would have to be revised. That revision followed in part because of the Wilde case, but it was also prepared by a broad investigation into homosexuality that had begun in the late 1860s.

No one doubted that homosexuality existed, but a debate raged in medical circles about its cause. Four explanations were offered. The most critical view was that homosexuality was a crime which resulted from a breakdown of the moral faculty of the human mind. Paul Moreau developed this view in 1880 when he classified it along with rape and necrophilia.[17] The second explanation, that homosexuality was a disease, dominated the medical literature and was elaborately explored by Tarnowsky, Krafft-Ebing, and Moll.[18] Krafft-Ebing argued that it generally resulted from masturbation and was unquestionably a sign of hereditary degeneration exacerbated by "perverse" environmental influences. A third and more accepting explanation was first articulated in 1868 by the German physician Karl Ulrichs, who argued that homosexuality was an anomaly—unusual but neither criminal nor pathological.[19] He offered a theoretical explanation which got much support among homosexuals until well into the twentieth century—that the sexual invert (he called them "Urnings") resulted from a female soul being trapped in a male body. He further argued that homosexuality was innate—not the consequence of bad habits or evil influences—and that most homosexuals were incapable of being converted into heterosexuals. They are, he insisted, neither morally, physically, nor intellectually inferior to others. In 1869 a Hungarian physician, Benkert, first used the term "homosexual," and in the following year Carl Westphal coined another term that quickly came into use—the sexual invert. In 1882 Charcot and Magnan communicated this theory to French medical circles with their article *"Inversion du sens génital,"* which regarded the phenomenon as an anomaly caused by sexual traumas.

When Krafft-Ebing devoted the largest part of his study of sexual pathology to an examination of homosexuality, he retarded the development of the more tolerant views that had been circulating for almost twenty years. His unrelenting condemnation of homosexuality as pathological, however, triggered a number of counter-arguments which generated a fourth view—that homosexuality was a virtue, a source of artistic inspiration and of sensitive love that runs stronger precisely because it is subject to so much misunderstanding and prejudice. A number of the most prominent defenders of this position were themselves homosexuals who were attempting to find intellectual support for the very instincts that threatened to make them outcasts.

Whitman had ventured a few verses in praise of "manly friendship," but he veiled any explicit reference to homosexuality.

Thus in "Spontaneous Me" he wrote: "The young man that wakes deep at night, the hot hand seeking to repress what would master him . . . ," but he did not reveal whether he was referring to auto-erotism or mutual masturbation. The final line of "Children of Adam": "Touch me, touch the palm of your hand to my body as I pass, / Be not afraid of my body" also did not explicitly refer to homosexuality. Many years later, when John Addington Symonds wrote to Whitman and asked whether "Calamus" was intended to praise homosexual love, Whitman rejected such interpretations as "morbid": they, he added, "are disavowed by me and seem damnable." [20]

Symonds belongs with the group of critics who regarded homosexual love as potentially superior to heterosexual love, but he equivocated by distinguishing between the promiscuous carnality of some inverts and the pure love between men that flourished among the ancient Greeks. Symonds accepted Ulrichs's explanation of the invert as a female soul in a male body, and he believed himself to be one. He set out his own views on the subject in *A Problem of Modern Ethics* in 1891, which concluded with a comprehensive criticism of the current prejudice against homosexuals.

> The points suggested for consideration are whether England is still justified in restricting the freedom of adult persons and rendering certain abnormal forms of sexuality criminal . . . after it has been shown (1) that abnormal inclinations are congenital, natural, and ineradicable in a large percentage of individuals; (2) that we tolerate sterile intercourse of various types between the two sexes; (3) that our legislation has not suppressed the immorality in question; (4) that the operation of the Code Napoleon for nearly a century has not increased this immorality in France; (5) that Italy, with the experience of the Code Napoleon to guide her, adopted its principles in 1889; (6) that the English penalities are rarely inflicted to their full extent; (7) that their existence encourages blackmailing, and their non-enforcement gives occasion for base political agitation; (8) that our higher education is in open contradiction to the spirit of our laws.[21]

The most popular study of homosexuality to appear around the turn of the century, Havelock Ellis's *Sexual Inversion*, incorporated much of Symonds's material and shared his view that homosexual love might be innate and its practice "natural and normal." After surveying the variety of causes of homosexuality (heredity, frustrated "normal" love, seduction, national propensities, class origins, certain professions), Ellis introduced another cause—the latent organic bisexuality of both sexes. This idea did not originate with Ellis, but he used it to explain homosexuality and generated a good deal of discussion among psychiatric circles.[22]

This theory was part of a general reevaluation of thought about sex which, as we have seen, was challenging the idea that men and women were polar opposites. Developments in hormone theory and genetics had begun to reveal that only the sex cells were entirely sex-typed and that the other body cells were bisexual. Hence the personalities of the sexes need not be characterized in sets of contrasting opposites.

Because speculation about normal human bisexuality was championed by homosexuals, the theory became tainted with much unfavorable prejudice. Magnus Hirschfeld founded a journal in 1899 to study "sexual intermediary stages"—physical and psychological hermaphrodites who provided visible bodily manifestations of human bisexuality. In the late 1890s Hirschfeld began to publish studies on homosexual love, in which he explained that homosexuality was a consequence of the original hermaphroditic disposition (*zwitterig Uranlage*) of mankind. In 1897 he founded in Berlin a society to establish a scientific study of homosexuality. In 1899 he began the *Jahrbuch für sexuelle Zwischenstufen* (Yearbook for Sexual Intermediary Stages), which provided the first regular public literary forum for the study of homosexuality. The journal spearheaded the campaign for repeal of Paragraph 175 of the German Penal Code and supported Hirschfeld's theory that the homosexual was someone in whom the bisexual disposition obscured the normal sex differentiation. Several articles showed photographs of hermaphrodites to illustrate this normal bisexual tendency.

In 1910 Hirschfeld explained transvestism as another manifestation of the fundamental bisexual nature of human existence. There is a continuity between masculinity and femininity—not an unbridgeable chasm. In the homosexual, the hermaphrodite, and the transvestite, the middle ground of human bisexuality overpowers the sex differentiation which leads most individuals to heterosexual roles. Hirschfeld was too keenly aware of the agony suffered by homosexuals from societal condemnation to argue that their love was superior to heterosexual love, but he led a vigorous campaign in its defense and contributed to the revision of sex roles that was under way at that time.

The most forthright arguments for the superiority of homosexual love were made by the English critic Edward Carpenter. Even Krafft-Ebing had conceded, Carpenter argued, that many homosexuals were "highly gifted in the fine arts, especially in music and poetry." The homosexual may be able to show society how to regulate its affections free from the "social and proprietary jealousy of modern marriage." If heterosexual relations were patterned after the freer interchange of the homosexual world, prostitution could be reduced. Carpenter concluded that homosexuals "are superior to the normal [?] men . . . in respect of their love-feeling, which is gentler, more sympathetic, more considerate, more of a matter of the heart and less one of mere physical satisfaction than that of ordinary men." [23] With this statement homosexuality began to shift from a purely defensive posture to an offensive, almost aggressive, assault on the heterosexual rule that had kept it in hiding for so long. That cause was led into the full light of day in the work of a leading French cultural figure, André Gide, who struggled for over thirty years before he overcame his fear of public exposure.

In January of 1895 Gide met Oscar Wilde in Algiers, a short time before Wilde's trials were to begin. At that time, Gide came to realize that his homosexual impulses constituted not an occasional deviation, but the heart of his sexual instinct. In the course of conversations with Wilde, Gide began to overcome his shame and look upon those impulses as a sign of aesthetic superiority. But he nonetheless succumbed to convention and married in October, 1895, under the influence of a physician who assured him that once he settled into a marriage he would forget his homosexual impulses. Gide quoted the extraordinary words used by the doctor: "You give me the impression of someone who is starved and who up until now has tried to subsist on pickles." [24] The marriage was never consummated, and his sex life consisted of surreptitious homosexual affairs. In an effort to deal with the destructive force of constant concealment and fear of discovery, he began to collect material on the history of homosexuality. His first literary exploration of the subject was a disguised autobiographical novel, *The Immoralist* (1902), in which a young man sets off with his new bride to visit Algiers, where his health breaks down. In the course of his recovery he reflects on the joys of the body and exclaims about the pleasure of feeling the return of the "rhythm of his muscles." He finally consummates the marriage, but soon begins to become fascinated with young boys. The homosexual attachments become more and more explicit, until a prostitute with whom he has settled after his wife dies tells him that she suspects that he prefers the boys to her. In the final line of the story, he concedes: "There may be some truth in what she says."

Gide began to think through a more forceful defense of homosexuality in 1911, when he distributed privately twelve copies of two dialogues he had written on the subject, entitled *C. R. D. N.* He distributed another twenty-one copies in 1920, and in 1924 finally published three complete dialogues under the title *Corydon*. Throughout his life he considered it to be his most important book.

The interlocutor in the dialogues visits Corydon to discuss a current trial on homosexuality. Corydon explains how he discovered that he was a homosexual. He was engaged, and his fiancée's brother declared his love for him and then killed himself. This tragic incident led Corydon to begin writing a book on homosexuality in which he developed the argument that love between men is natural and superior to heterosexual love. He gleaned supporting evidence from a variety of biological, psychological, and historical theories. Nature produces more males than females, he argued, and they are also more beautiful: hence they are attracted to each other. Females are drab and often need nothing more than their smell to attract males. Women do not have to be beautiful; "it is sufficient that they be in good odor." [25] Corydon then eulogized male homosexuality. "Each renaissance or period of great artistic activity is always accompanied by a great outbreak of homosexuality. . . . And when the day comes to write a history of homosexuality, it will be seen to flourish not during periods of decadence, but during those glorious, healthy periods, when art is most spontaneous and least artificial." [26] Corydon's preachings in the dialogues are bold, but Gide was reluctant to have them published openly until 1924.

By that time he was preparing to abandon the thin fictional disguise that he had previously used and make his confession publicly in the first person. The appearance of his autobiographical essay, *Si le grain ne meurt* (If It Die) in 1926, marks the end of an epoch in the history of homosexuality. H. Stuart Hughes has speculated that then "for the first time . . . since classical antiquity, a public figure in the Western world had openly admitted his 'forbidden' desires." [27]

Why was it such a struggle for Gide to make public his homosexuality? Why *is* homosexuality so threatening to heterosexuals and such a source of shame for those who engage in it? The answers are implicit in the history I am tracing. Throughout the nineteenth century, society prescribed an elaborate set of rules about how the human body was to be understood and experienced. Any deviations were particularly threatening. Mutual masturbation, fellatio, or anal intercourse seemed to most moralists to be "unnatural" acts—a sign of criminal tendencies, moral depravity, mental disease, or bad taste. There was some progress in toleration away from the legal prohibitions and the moral taboos, but even the most independent defenders of homosexuality could not liberate themselves from the feeling that something was wrong with it. It is a telling comment on the history of homosexuality of this period that its most outspoken defenders either ignored female homosexuality or regarded it, as did Gide, as inferior to male homosexuality. In 1926 European homosexuals still had a long way to go before they could achieve a return of a golden age of homosexual freedom such as they believed had existed in ancient Greece.

Chapter 13

What a Young Girl Would Know Who Knew What a Young Girl Ought to Know

In tracing the histories of the subjects covered in this book, one is vulnerable to a tempting methodological flaw—to focus generally on the most outrageous ideas about human sexuality in a given age while ignoring the more reasonable ideas and practices of the period. Studies of Victorian sexuality are particularly prone to such distortions, as some recent literature has argued.[1] A good case could be made that the Victorians' sex ethic was reasonable in light of the dangers of venereal disease and the limited knowledge of and access to contraceptive techniques. Sexual restraint, even rejection of men, gave women of the period a measure of freedom from endless pregnancies and the inexorable demands of motherhood. I have perhaps dwelled on some of the more shocking practices of the age to illustrate the fanatic extremes that the denial of sexuality produced, although my general argument is that, from the mid-nineteenth century, Europeans and Americans began to grope for a fuller understanding of their corporeal existence. I believe I am justified in giving in to the temptation to present a parade of horrors, because however advanced the movement toward a more rational sex ethic was, it was not until the First World War that European culture began radically to change its sex ethic. I begin with one "idea" about sex education which, I believe, would not have been suggested after that war and after the spread of psychoanalytic thought.

In G. Stanley Hall's *Adolescence* (1904) we read: "Some think, at least for girls, all that is needed can be taught by means of flowers and their fertilization, and that mature years will bring insight enough to apply it all to human life"—a typical statement for the age. But then Hall continues: "Others would demonstrate on the cadaver so that in the presence of death knowledge may be given without passion. This I once saw in Paris, but cannot commend for general use." [2] I know of no more solid evidence of the incapacity to deal with human sexuality than this suggestion that the facts of life be taught by use of a dead body. This thought was not typical for the age, but it ought to be remembered as a characteristic, though exaggerated, product of it. It appears on the eve of the revolution in thought about the human body that was anticipated in the studies of sex that began to appear in the 1880s and which then attained a more forceful expression in the work of Freud. Let us now trace some of the prehistory of this chilling pedagogical gem by reviewing developments in contraceptive techniques that generated a continuing debate over sex education in the nineteenth century.

There are two aspects to contraception: it is an effort to control the sexual impulse either by regulating the time of sexual activity or by cluttering it up with various appliances or substances, but it is also a means for separating the procreative function from sexuality, allowing sex to be enjoyed for itself. Throughout the nineteenth century these two views colored an endless debate over the moral and aesthetic value of contraception. Defenders hailed it as a release from the anxiety that hindered complete sexual pleasure, while its detractors thought it would "degrade the finest moral instincts" and produce a "bestial sensuality." [3] Many feminists feared that contraception would merely permit the free indulgence of male sensuality, which had been at least partially controlled by the possibility of conception.

The man who wrote what was to become the keynote address for the birth control movement in the nineteenth century, Thomas Malthus, explicitly argued against the use of contraception. He believed that sex was to be engaged in only for procreative purposes and therefore recommended the postponement of marriage as "moral restraint" to control population.

At the beginning of the nineteenth century there were two kinds of mechanical contraceptive devices: condoms made from fish skins or animal bladders, and various kinds of sponges that were inserted into the vagina. That latter method was recommended by the English reformer Francis Place in 1822, although he also mentioned as another possibility *coitus interruptus*.[4] Charles Knowlton's *Fruits of Philosophy* (1832) popularized the method of douching, and then in 1838 the German gynecologist Friedrich Wilde introduced yet another method—a rudimentary diaphragm called the "cervical rubber cap," which was molded to fit over the cervix and removed during menses. This device was further improved with the "occlusive pessary," developed by a German anatomist named Mensinga. The Mensinga pessary began to come into use in England following its strong recommendation in Dr. H. A. Allbutt's *The Wife's Handbook* (1887). Allbutt also suggested other methods such as condoms, diaphragms, sponges, injections, *coitus interruptus*, and rhythm. The discovery of the process of the vulcanization of rubber in 1843 made possible the production of rubber condoms to replace those made of animal skins, but they did not begin to be distributed in large numbers until the 1870s.[5]

Throughout the first half of the nineteenth century, calculations of the "safe period" were often wrong because physicians tended to think that humans, like most other mammals, were fertile during menses. It was only in the 1860s, when discoveries on the nature of human sexual reproduction revealed that the woman is generally fertile around the middle of her menstrual cycle, that rhythm methods began to be reliable, although erroneous theories about safe periods continued to be published until the end of the century.

By the 1880s Europeans had a variety of contraceptive techniques to choose from, but the most popular technique continued to be *coitus interruptus*, a method which required an enormous level of self-restraint and allowed the bourgeois man to demonstrate his command of some cardinal middle-class virtues—responsibility, prudence, and self-control. The high point of this sexual ethic was achieved by John Humphrey Noyes, founder of the Oneida Community in New York in 1848. To spare his wife future misery following her disappointment from four stillborn children, he developed a technique of sexual intercourse which did not culminate in orgasm and ejaculation. He explained his method in *Male Continence* (1872), in which he insisted that the sexual process is involuntary and "can be stopped at any moment."[6]

Around the turn of the century some neurologists began to suspect that *coitus interruptus* might be causing a number of nervous disorders. One German physician warned that it could cause impotence, spermatorrhea, swelling of the prostate, *"asthma sexualis,"* chronic diarrhea, insomnia, or nightly pollutions.[7] During the 1890s Freud believed that some neuroses were caused by *coitus interruptus* in accord with his theory that certain "actual neuroses" were a direct reaction to sexual frustration. Iwan Bloch tried to dispel the alarm in 1907 when he advised that the method was not so damaging as was previously believed, although he did continue to advise against one method which some still believed prevented conception—prolonging intercourse. He reported that some individuals had stretched it out long enough to smoke and read during the sex act.[8]

The manuals on sex education for men devoted inordinately large space to the dangers of the uncontrolled sexual impulse, and particularly the dangers of masturbation. Let us look at one typical manual in detail. An American physician, George H. Napheys, published *The Transmission of Life: Counsels on the Nature and Hygiene of the Masculine Function* in 1878. It was intended to instill caution and self-restraint among its readers, who were reminded of the endless dangers from their sexual instinct. The period of virility for American men, he argued, was between twenty-five and forty-five. Premature sexual activity, in particular masturbation, would cause an earlier cessation of the sexual impulse in adulthood. The sexual instinct should generally be stimulated as little as possible to preserve health and ensure adequate potency when it is needed. He then outlined the scenario of progressive deterioration for the reckless. Masturbation leads to nocturnal emissions, which in turn may lead to chronic and involuntary emission of semen—spermatorrhea. With a man's sexual energy dripping out in this wasteful manner, his adult sex life would be shortened and imperfect. Napheys further counseled young men that they might contract gonorrhoea by having sex with a woman during her menstrual period. One young man committed suicide after he had contracted gonorrhoea during his honeymoon, reasoning that his wife must have been unchaste. Napheys offered a substitute explanation—he contracted gonorrhea from her because they had sex during her menstrual period. Young men ought not to expect that their wedding night has to be a "bloody rite," and "in stout blondes it is even the exception rather than the rule." [9] He then considered sexual passion in women. "A vulgar opinion prevails that they are creatures of like passions with ourselves; that they experience desires as ardent, and often as ungovernable, as those which lead to so much evil in our sex. Vicious writers, brutal and ignorant men, and some shameless women combine to favor and extend this opinion. Nothing is more utterly untrue." [10] Napheys then cited the by-then famous passage from William Acton that only seduced or degenerate women experienced sexual excitation. He added a warning that indulging in intercourse during pregnancy would surely injure the fetus and somehow contaminate the mother's milk.

Napheys's manual was typical of its age. Historians have perhaps dwelled excessively on these unusual ideas, but they were repeated endlessly until well into the twentieth century. One of the most famous handbooks for young boys published in Germany, Hans Wegener's *Wir jungen Manner* (1906), greeted the curious adolescent on the title page with a drawing of a lion killing a snake. The snake of human sexuality could only be extirpated by the most constant and zealous means. The instructions to young men were explicit and frightening, but those for the young girl were even more destructive, because behind the rhetoric about birds and flowers was the unmistakable message that any manifestations of her sexual impulses were abnormal. The young man could at least reassure himself that if he carefully regulated his sexuality he could be a decent individual, but the little girl had to repress even the faintest stirrings of her sexuality from the outset. No wonder it manifested itself so frequently in vertigo and hysteria.

I have taken the title for this chapter from *What a Young Girl Ought to Know*, published by Mary Wood-Allen and Sylvanus Stall in 1897. The sex education contained in this book is the most destructive piece of moralistic misinformation that I have found. The message is at best confusing. After beating around the bush for one hundred pages about flowers, insects, and birds, the authors get to the heart of the matter. They approach it with the strangest analogy I have ever come across.

> You would not put sticks or stones in your ears nor let any one else do so. Every organ of the body is sacred and should be protected, and this is just as true of the sexual organs as of the eyes or ears. You should never handle them or allow any one else. And yet, girls sometimes form a habit of handling their sexual organs because they find a certain pleasure in so doing. . . . It is called the solitary vice. . . . It leaves a mark upon the face so that those who are wise may know what the girl is doing. . . . We can almost always tell when a girl begins this habit of solitary vice . . . she will soon become peevish, irritable, morose and disobedient. . . . She may become bold in her manner instead of being modest, as a little girl should be. She will manifest an unnatural appetite, sometimes desiring mustard, pepper, vinegar, and spices . . . which appetites certainly are not natural for little girls.[11]

The little girl then gets some instruction on how to care for other parts of the body. Again, the rhetoric is insipid.

> There are certain organs of the body which we use openly in society; there are others which are to be used only in solitude, not because they are vile but because it is refined and polite not to use them in public. We do not carry our garbage pail into the parlor, and yet we know it is just as important that the garbage pail should be emptied every day. . . . So there are certain offices of various organs of the body which are performed by polite and delicate people only in solitude. . . . A certain part of our food remains as waste and must be cast out of the body. . . . It is very important that this bodily housekeeping shall be promptly and regularly attended every day.[12]

The little girl is then warned about the wicked influences of other children.

> Children sometimes go with each other to the closet and often their talk is not what it should be. The little girl

> who values her modesty . . . will never allow anyone to talk to her concerning any part of her body in a way that is not sweet and pure, and if any child ventures to give her information concerning herself that seems to her such as she would not tell her mother, the wisest thing for her to do would be to say: "I would rather you would not tell me about it. I will ask my mother. . . . Mother tells me everything that I ought to know and she tells it to me in such a way that makes it very sweet to me, and so I have my little secrets with mother, and not with other girls." [13]

The authors then introduce God.

> Your body is a beautiful house in which you dwell, and not only that but . . . it is a sacred temple in which God dwells with you.

By now it had to be getting confusing.

> A temple, you know, is a place of worship, and it is the same as if He called the body a church. We feel that we must behave very properly when in church.[14]

They again return to the intimate relation between a girl's body and God.

> Evil thoughts create actual poisons in the blood [and] all kind and good thoughts create life-giving forces in the blood. So if we want to keep this temple of the body in good condition, we want to invite into it as guests, only the sweet and lovely thoughts which can truly be associated with God himself. He has promised to dwell in us and He cannot dwell in perfection where there are evil thoughts or feelings.[15]

Let us consider what a young girl would learn from reading this book. The body is a temple in which God dwells, and she should not allow unwelcome guests into it. The authors explicitly say that she ought not to allow sticks or stones in her ears or bad ideas or feelings into her blood, but all that is obviously a disguised instruction not to allow any part of the male body into her vagina. Any trespass will leave her forever contaminated. She would further understand that her bowels are not to be "used openly in society" but "only in solitude." This regular "housekeeping" is important to prevent worms from gathering in her bowels, wandering out, causing local irritation, and thus stimulating her to scratch herself and lead to the "solitary vice." Furthermore she ought not to discuss her body with anybody but mother, who will make everything "very sweet." She should work to suppress evil thoughts, which create evil blood and can even infect the atmosphere around her. Above all she should keep the temple which is her body clean in thought and deed, because God dwells in it and does not like to dwell where there is evil.

The book offers no discussion of her sexual feelings and no explanation of how she might achieve some degree of sexual pleasure. There is no discussion of venereal disease beyond some vague references to contamination that can result from any sexual contact, and there is no mention of any method of birth control. The only source for further information she is advised to seek is from mother. But the Victorian mother-daughter relationship did not always provide for the happy instruction in the facts of life that this book recommended. Frank Wedekind, in his play on the problem of sex education, *Spring's Awakening* (1890), reconstructed what he considered to be the more common anxiety-ridden manner in which sexual information was disseminated. In the following scene Wendla Bergmann is begging her mother to tell her how children come into the world.

> MRS. BERGMANN: It's enough to drive one crazy! Come, child, come here, I'll tell you! I'll tell you everything. . . . Merciful Providence! Only not today, Wendla! Tomorrow, the next day, next week . . . whenever you want, Sweetheart . . .
> WENDLA: Tell me now, Mother; tell me now! Right now! Now that I've seen you so upset, *I* can't calm down.
> MRS. BERGMANN: I can't, Wendla.
> WENDLA: Why not, Mother? Look, I'll kneel at your feet and lay my head in your lap. You can put your

WHAT A YOUNG GIRL WOULD KNOW

apron over my head and talk and talk, just as if you were completely alone in the room. I won't flinch; I won't cry out; I'll listen patiently, no matter what happens.

MRS. BERGMANN: Heavens knows, it's not my fault, Wendla. In God's name! Come child, I will tell you how you came into this world. Are you listening, Wendla?

WENDLA *(under the apron):* I'm listening.

MRS. BERGMANN: It won't work, child! I can't take the responsibility. I deserve to be thrown in prison—they should take you away from me.

WENDLA *(under the apron):* Courage, Mother!

MRS. BERGMANN: All right then, listen . . .

WENDLA *(under the apron, trembling):* O God, O God!

MRS. BERGMANN: To have a child—do you understand me, Wendla?

WENDLA: Quick, Mother—I can't stand it any longer.

MRS. BERGMANN: To have a child—one must love the man—to whom one is married . . . love—love him, I tell you—as one can only love one's husband! One must love him with one's whole heart, so much . . . so much that one can't even express it. One must love him, Wendla, as you at your age are incapable of loving. . . . Now you know.

WENDLA *(getting up):* Great—God—in Heaven!

MRS. BERGMANN: Now you know what trials lie before you! [16]

The sex education that Wendla acquires, trembling, with her head in her mother's lap and with an apron over her head, is worse than none at all, because she takes her mother literally and concludes that she cannot have a child because she is incapable of loving a man that way and has to be married first. She seduces a young man, becomes pregnant, and succumbs to "Mother Schmidt's Abortion Pills."

Sex manuals for American women of the Victorian era expressed a number of confusing and anxiety-provoking theories. Some said that impure thoughts would stimulate excessive growth of the sex organs and lead to a corresponding increase in the sexual appetite.[17] One moralist counseled that "the practice of novel reading is one of the greatest causes of uterine disease in young women."[18] The intimate and delicate connection between the young girl's mind and her body, particularly her sex organs, was further emphasized by another manual which warned that if a girl is jilted by the first man she loves, her "female functions and organs" will be forever crippled. This will further "break down the very elements of female attractiveness, because of the perfect reciprocity which exists between the mental and physical sexuality."[19] Again a double bind is imposed upon female sexuality—it does not exist, but exerts a life-and-death hold over emotions.

By the 1890s, however, these ideas began to give way to more positive counseling about both male and female sexuality, and by the First World War one could speak of a sex education movement in Europe and America.

A Swedish professor, Seved Rebbing, gave a series of lectures on sex in 1886 which criticized medical books that emphasized the dangers and diseases of sex.[20] Although he concluded his lectures with a comprehensive survey of venereal diseases, he noted that fear of them need not be the dominant theme of studies of sex. He warned against excessive indulgence in sex and set an average limit of three to four times per week, particularly during the early stages of a marriage in order to cultivate a "sensitive" relationship. In 1889 the Scottish biologist Patrick Geddes published *The Evolution of Sex*, which argued for traditional male-female roles from a biological presupposition that females were metabolically conservative, while males tended to dissipate energy. Geddes also introduced a plea for the acceptance of female sexuality, although he did support the traditional notion that women ought to conform to their biological fate and accept the passive-receptive role. But the book was most important for its pathbreaking tone of calm and thoughtfulness, its systematic and sober discussion of the sex organs and their proper function.

The French public was introduced to the latest theories about sex in a series of books by Pierre Garnier, which were intended to facilitate sexual relations, though not to the point of imprudence or overindulgence.[21] German women got access to a popular handbook of 970 pages on feminine hygiene in 1901 from the physician Anna Fischer-Dückelmann. By 1908 it had sold 500,000 copies and had been translated into ten languages. She explained that her book was intended to counter the kind of thinking behind the complaint that art revealed too much of the naked body. She wanted to liberate women from the prejudice which had so degraded the study of woman's natural instincts. The human body is not intrinsically terrifying, but has become so because it has been estranged from us. German women, she said, knew very little about their bodies, and so the first chapter, complete with illustrated paper fold-out drawings of the female body, offered instruction on human anatomy. A series of drawings which showed the embryological development of the male sex organs out of the female was to help correct the prevalent idea that men and women were completely opposite in their sex and temperament. She commented critically on the German term for the female genitals, *"Scham,"* which also meant shame. She then explained that the nerves in the clitoris were connected to the spinal cord and the brain, and therefore women ought to be able to experience sexual excitation from images or thoughts. But, she speculated, many women completely lacked this kind of stimulation. Her book would presumably help women to overcome this deficiency. An argument for clothing reform urged that the body be allowed to breathe, exercise, and see sunlight. "Today with the growing recognition of the real meaning of the skin, mankind is beginning to discover the wonderful effect of *Lichtluftbäder.*" [22] "Beauty is power," especially for women, she wrote, and a large section of the book offered beauty hints. Also included was a list of feminine features highly valued at that time, such as a delicate bone structure, a full breast, broad hips, long full hair, but sparse and light hair on the body generally, in the armpits, and in the pubic region.

The counseling on sex life was contradictory. Fischer-Dückelmann reminded her readers that this chapter on female sex life was written with the best of intentions and ought not to be misinterpreted. At times she concurred with the traditional negative evaluation of female sexuality, arguing that the more highly developed a woman's intellect, morality, or class status the less developed was her sexuality. But she then insisted that women must not have a hidden and wasted sex life. Her explicit instructions for the sex act were daring. "The chances of conception are better if the uterus is in a favorable position and if the woman does not cooperate with muscular activity." [23] Once again women were told that sexual feelings and sexual activity would endanger their highest function—the conception and bearing of children. Nevertheless women were advised to give up the excessive modesty that kept them from learning about sex and begin to take some initiative. A completely ignorant bride, it was warned, would only provoke the instinctual bestiality of her man by confusing him and by not responding in a way that was based on some understanding of her own sexual capacity and needs. She was emphatic about the existence of female sexuality: "One should not think that women do not also possess sexual desire." [24] Energetic and powerful women, she noted, were well endowed with a sex instinct and suffered if not married, and no woman should be ashamed of her natural feelings. Her message to women was ambiguous, however, because she continually shifted back and forth between urging women to enjoy their sexuality and cautioning them against it. After a good deal of positive counseling she added that many good marriages were ruined just because the woman abandoned her shame. This "shamelessness" disgusts men. Couples ought to have sex once or twice a month to leave strength for higher pursuits and to ensure strong children. If necessary the couples should avoid sleeping in the same bed, or even in the same room, if they cannot control their sexual impulses. She recommended separate beds against opposite walls and screened by a curtain if necessary to curb the sexual appetite.

A careful reading of this manual reveals that the overall message on female sexuality is confusing. But the historical impact was not, because, I would speculate, the women of this period did not read it carefully. They dwelled on the positive advice about their sexual needs and tended to overlook the traditional moralizing that was added to give the overall impression of a restrained text. I believe that her book was misinterpreted and, further, that she intended it to be, in spite of her disclaimer to the contrary.

Around the turn of the century, a number of popular studies of sex descended on the German public, beginning in 1905 with the most famous, August Forel's *The Sexual Question.* Forel argued far more forthrightly than did Fischer-Dückelmann that human sexuality was to be revered and cultivated. There was no ambiguity in his initial pronouncement: "Sexual life is beautiful as well as good." He went on to condone abortion for victims of rape, advocate sex education for children, and champion the emancipation of women. He concluded with an ominous section that warned of the progressive deterioration of the Aryan race and he advised his readers about a number of eugenic practices which might control that progressive deterioration. The book achieved enormous popularity largely because of its discussion of sexual problems and their solutions. While Forel defended monogamy, a German philosophy professor, Christian von Ehrenfels, attacked the institution in *Sexualethik* (1907). Only a polygamous society, he argued, could permit the "virile discharge" of sexual substances which was requisite for a fulfilling sex life. In the same year Bloch's *Sexual Life of Our Time* commented on the progress of the study of sex and the impact of various media on it. Bloch noted that newspaper accounts of secret affairs, suicides, rapes, and sex crimes were forcing sex life into public attention more than ever;[25] even the advertising of charlatans offering to cure venereal diseases was promoting public discussion of sex. The progress of sex education moved on its negative side as well among certain circles through scandal, blackmail, and quackery.

The revelations of psychoanalysis provided a rich new source of ideas on sex in the early years of the century, and Freud explicitly commented on the impact of sex education on the mental health of a culture. An unmistakable implication of all of his work on the neuroses caused by sexual inhibition, sexual frustration, or sexual excess was that a more relaxed sexual ethic and broader dissemination of sexual knowledge would improve mental health. In " 'Civilized' Sexual Morality and Modern Nervous Illness" (1908) he decried the pathogenic effects of sexual ignorance. But in his *Three Essays on Sexual Theory* (1905) he offered a contrary theory that parents must preserve a certain measure of mystery about sexual matters with their children in order to stimulate their first little research projects about why the sex organs of boys and girls are different, what Mommy and Daddy do in bed at night, and how babies come into the world. The prototype for all future education, Freud believed, was this first sexually motivated research into the mysteries of human sexuality.

In the next two chapters I will discuss important sources for the "sexual revolution" of the early twentieth century—the First World War and the work of Freud. But to conclude this chapter I shall jump ahead of my story to consider a book on sex education by a Dutch gynecologist which illustrates the changes in sexual morality that were in the making.

Theodore Van de Velde's *Ideal Marriage* was first published in Dutch and German in 1926 and was soon translated into French and English. Van de Velde began with the candid statement that good marriages must be maintained by a lively and imaginative sexual life. Although much of the book concerned information on female sexuality, he addressed himself to married men because "they are naturally educators and initiators of their wives in sexual matters." But they often do not realize their own limitations and blame all failures on their wives, whom they dismiss as frigid to justify their own marital infidelity. These husbands should cultivate sexual happiness by learning to achieve "vigorous and harmonious" sexual relations. Most men, for example, do not know that women normally take longer to become sexually excited. The typical European man "cannot understand why the Hindoo women, used to the sexual assiduity and skill of their own men, mock the clumsy Europeans as 'village cocks.'" His positive counseling was revolutionary. Certain individuals' sense of smell has a particularly high erotic potential. These "olfactory types" should not suppress this aspect of their sexual life, but should "become acutely conscious of the enjoyment they derive from the subtle and various scents exhaled by the body they love." Following a long and detailed account of the physiology of the male and female sex organs, Van de Velde offered some instructions for maximizing sexual pleasure. This instruction derived from a premise about human sexual anatomy which until that time had never been made so explicit—that men and women were not ideally constructed to maximize each other's pleasure during intercourse. Women become sexually aroused more slowly than men, and therefore some provision must be made to arouse them sufficiently to have an orgasm during the sex act. And to complicate this procedure, the focus of women's sexuality, the clitoris, is often not placed properly or is not large enough to be stimulated during intercourse. The erect penis is supposed to stimulate the clitoris during intercourse, "but unfortunately, often and perhaps even generally this does not happen in modern women of our race; partly, or principally, because of insufficient development of the clitoris, of its relatively high position on the symphsis, and because of very slight pelvic inclination." This evaluation of the racial inferiority of European women was most unusual at this time, but it underlined the urgent need for men and women to attend to the problems created by the organic incompatability of the sexes.

To counter this tragedy of human anatomy Van de Velde suggested some techniques which were to make the book famous. To prepare adequately for sexual intercourse the couple must engage in love play. This "prelude" is important not only to ensure that the woman enjoys herself during the sex act, but also to prevent physical injury which can result from too rapid or too vigorous movement by the man when a woman is not properly lubricated or relaxed. The "erotic kiss" may involve *"Mariachinage,"* in which the couple explores and caresses the interior of each other's mouths with their tongues. The prelude may also involve the use of teeth in "nibbling" gently at one another around the lips, neck, and ears. Kissing may wander even further from the traditional anatomical terrain and explore other "erogenous areas" of the body from the fingertips to the genitals. The most controversial part of the book was his recommendation of the "genital kiss," although he emphatically condemned the practice of oral-genital sexuality up to the point of orgasm. It is only recommended to help arouse each partner in preparation for the achievement of orgasm during sexual intercourse.

Van de Velde described the artful approach to the genitals by the hand, before proceeding to the genital kiss. "After gentle strokings and claspings of the accessory organs, the hand should lightly and timidly brush the abdomen, the mons-pubis, the inner side of the thigh; alight swiftly on the sexual organs and pass at once to the other thigh. Only by a cautious and circuitous route should it approach the holy place of sex and tenderly seek admittance." [26] This kind of graphic instruction was substantially different from even the most enlightened books on sex education that had appeared before the war. The early pioneers of sex reform were too deeply embroiled in challenging the laws and fears and repressions of that earlier period to embark on the kind of genuinely positive counseling in the art of love-making that one finds in *Ideal Marriage*. Of course these instructions were always available in certain exotic books such as the *Kama Sutra*, but they never got the kind of attention that Europeans gave to them following this pioneer work in sex education. The free literary exploration of human sexuality that D. H. Lawrence included in *Lady Chatterley's Lover* was facilitated by the new attitudes about sexuality that Van de Velde and the movement for sex reform had helped to generate.

If a young girl consulted *Ideal Marriage* she would at least learn enough to have a fighting chance for a sexually gratifying marriage. It is difficult to imagine how this would have been possible thirty years earlier.

Chapter *14*

*The Discovery of the
Meaning of the Body*

The title of my book is a variation of the phrase "Anatomy is destiny," which Freud used in 1912 and then again in 1924 to explain the way the anatomy of human sexuality determines its nature. The first reference underlined Freud's theory that the sadistic component of the sex impulse derives from the proximity of the sex organs and the excremental organs. "The excremental is all too intimately and inseparably bound up with the sexual; the position of the genitals—*inter urinas et faeces*—remains the decisive and unchanging factor. One might say here, varying a well-known saying of the Great Napoleon: 'Anatomy is destiny.' "[1] Although man has assumed an upright posture and the "upper layers" of the erotic instinct have turned away from their animal origins, the genitals remain "between urine and feces," and sex has retained its anal-sadistic character. The second quotation of this phrase emphasized the way female anatomy destines women to experience their "phallic" sexual life differently from men. "The little girl's clitoris behaves just like a penis to begin with; but when she makes a comparison with a playfellow of the other sex, she perceives that she has 'come off badly' and she feels this as a wrong done to her and as a ground for inferiority." Freud generalized about this phase of the development of female sexuality by arguing that "the morphological distinction is bound to find expression in differences of psychical development." He concluded with the terse reformulation of this theory—"Anatomy is destiny." [2] Accordingly, the essence of female sexuality derives from the young girl's first perception of the anatomical focal point of her auto-erotic masturbatory sexual excitation in an inferior version of the penis.

There has been a great deal of discussion about Freud's theory that female sexuality is derived from the structure and function of the sex organs, but that controversy ought not to obscure understanding of the historical significance of the *kind* of reasoning it involves, because that reasoning, I shall insist, constitutes a most important contribution to European thought. The ideas about human corporeality that had appeared in the course of the nineteenth century either as passionate declamation or passing comment are given a systematic and graphic presentation for the first time in the work of Freud. Critics and disciples have repeatedly debated his interpretation of the way human anatomy, in particular sexual anatomy, influences human existence, and in doing so they have all implicitly conceded that anatomy does profoundly affect the life of man. This theory was not entirely new with Freud, but he developed it in greater detail than had any of his predecessors and presented it in a form that was convincing, useful, and provocative. It convinced large numbers of psychiatrists and has been useful to the thousands of psychoanalysts who have since applied it to the interpretation of their patients' dreams and symptoms. The application of the theory to explain female sexual character was plausible enough to provoke a shower of critical attacks which have become particularly bitter during the past five years. "Anatomy is destiny" challenges the significance of the environment and of the mind in shaping human affairs. It does exaggerate the importance of the body by suggesting that it alone determines destiny, but the historical record shows that it was a useful distortion of the truth, one by which Freud no doubt intended to draw attention to an aspect of life which scholars until that time had tended to neglect.

The title of this chapter also exaggerates somewhat the originality of Freud's contribution to the age-old subject of the relation of the mind and body, but I believe the exaggeration is allowable because it focuses attention on the striking originality of Freud's work. Others before Freud had speculated on the relation of the body and the mind, and I have surveyed some of that work in the preceding chapters. But his work so far surpassed that of his predecessors in the clarity and the depth of his interpretations that it deserves being characterized as a discovery in the history of thought. To substantiate my interpretation of his contribution as a discovery of the meaning of the body, I will survey four kinds of mind-body interactions that Freud explored: the influence of the mind on the body in the formation of hysterical symptoms, the symbolization of bodily structures and functions in dreams and symptoms, the influence of the body in determining personality traits in the course of normal psychosexual development, and the way "civilized" morality imposes itself on the body by requiring instinctual renunciation and sexual repression.

CONVERSION SYMPTOMS

Freud's first serious professional aspirations began in biology. He shifted to private practice in neurology as an alternative career because he was advised that it would be impossible for him to secure an academic appointment at the University of Vienna. His first patients suffered from severe mental disturbances which often involved overt bodily manifestations, and his first important contribution to psychiatric thinking was a theory of the origins of certain kinds of somatic symptoms in hysterical patients.

In the winter of 1885–1886 Freud journeyed to Paris to study with the eminent French neurologist Jean Charcot, who was using hypnosis to diagnose and treat hysterical patients. Charcot was able to induce or remove certain bodily manifestations of his patients' mental disturbances by hypnotic suggestion. For five months Freud observed Charcot manipulate the paralyses, tremors, and nervous tics of his patients by commanding them to appear or disappear while they were under hypnosis. This illustration of the malleability of bodily symptoms under the influence of the physician's directives gave Freud a rich supply of material with which to develop his own theory of the relation of the mind and body. His first publication deriving from this experience was an article, written in 1888, though not published until 1893, which contrasted organic and hysterical paralyses. A hysterical paralysis, he argued, behaves "as though anatomy did not exist or as though it had no knowledge of it." For example, the hysterical paralysis of an arm could not be caused by any organic means, because no lesion in the nervous system could possibly produce it. Rather, "the lesion in hysterical paralysis will be an alteration of the *conception,* the *idea,* of the arm." [3] The mind dissociates the arm from the normal channel of associations and makes the body behave as if the arm did not exist. The result is a hysterical paralysis.

In 1894 Freud proposed a term to describe the somatic response to psychic trauma. "In hysteria the unbearable idea is rendered innocuous by its *sum of excitation* being *transformed into something somatic.* For this I should like to propose the name of *conversion.*" [4] The unbearable idea, he added, most frequently comes from the patient's sex life. The theory that certain ideas have a *quantity* of excitation associated with them introduced the economic aspect of psychoanalytic theory. Freud repeatedly used analogies of electrical and hydraulic systems to help him explain how ideas may have a certain quantitative value, a sum of psychic energy, attached to them. In this article of 1894 he first introduced the electrical analogy. "In mental functions something is to be distinguished—a quota of affect or a sum of excitation—which possesses all the characteristics of a quantity (though we have no means of measuring it), which is capable of increase, diminution, displacement and discharge, and which is spread over the memory traces of ideas as an electric charge is spread over the surface of a body." [5] All further discussion of hypothetical mental processes, such as cathexis, damming of libido, catharsis, or abreaction, hinged on this speculative and contradictory definition: that the energy which activates human behavior possesses all the characteristics of a measurable quantity except one—it cannot be measured. In 1909 he compared the flow and discharge of psychic energy to a hydraulic system. "When the bed of a stream is divided into two channels, then, if the current in one of them is brought up against an obstacle, the other will at once be overfilled." [6] The electricity and water metaphors gave graphic representation to his account of the fate of the sexual energy that he believed drove the entire psychic apparatus. Controlling this distribution of libidinal energy was the "principle of constancy," which attempted "to keep the quantity of excitation present as low as possible, or at least as constant." [7] With this scientific-sounding terminology Freud tried to bridge the gap conceptually between mind and body. Psychic processes could not be measured by specific energy values, but they nevertheless had some kind of magnitude which enabled them to modify bodily structures and processes like dammed water breaking through a new channel or electricity shorting out a circuit. The erection of the penis was an example of the normal action of mental processes on bodily states. Thoughts and images generate sexual excitation in the body, which stimulates the flow of blood into the tumescent penis until the system is overloaded and the nervous system shorts out in an orgasm as water behind a dam eventually bursts through. The erect penis and its orgasmic discharge became, for Freud, the quintessential symbol of sexual excitation, a graphic model of mind-body interaction. By the time of the publication of *Studies in Hysteria* in 1895, Freud had developed an elaborate theoretical apparatus to explain the meaningful connections between conversion symptoms and their traumatic origins.

In 1889 Freud began treating Frau Emmy von N., who exhibited some conversion symptoms—a pained facial expression, a tic in the face and neck, and intermittent "clacking" sounds. By way of introduction to his interpretation of these symptoms Freud repeated his theory of the dynamics of the psychic energy involved in the formation of conversion symptoms. "It is impossible any longer at this point to avoid introducing the idea of quantities (even though not measurable ones). We must regard the process as though a sum of excitation impinging on the nervous system is transformed into chronic symptoms in so far as it has not been employed for external action in proportion to its amount." [8] Freud interpreted the symptoms in accord with this theory of conversion and related each of them to some traumatic experience. Her pained facial expression was a somatic representation of the memory of the pain she witnessed and experienced during the many months she spent sick-nursing. The tic was related in origin to the clacking sound. During long hours of nursing her sick child when she had to keep silent to allow the child to sleep, she had developed an "antithetic idea" of making noise. The clacking and the tic were efforts to resolve contradictory impulses to remain silent and to make noise, to keep still and to move. The bodily sounds and movements were thus traced to their origin as a compromise between contrasting impulses.

In 1892 Freud was consulted about Fräulein Elizabeth von R., who had pains in her legs and experienced difficulty in walking, although she had no discoverable organic disorders. Like Emmy von N. she had spent much time nursing ailing members of her family, and Freud related the pain in her leg to the nursing experience. The particular spot on her thigh where the pain was most acute, she finally recalled, was the exact place where her father used to rest his leg while she bandaged it every morning. In the course of her analysis her painful legs began to "join in the conversation." If she was free of pain at the beginning of an analytic session, the pain would begin to mount as Freud pressed her to recall painful memories. It reached a climax as she related the most important part of the memory and then disappeared when the cathartic recollection ended. Freud thus came to use the pain as a "compass" to guide him to the most important corners of her past. The other symptom he traced to a series of painful experiences that had occurred while she was standing, walking, or sitting and which came to manifest themselves as an impediment to her walking ability. She gave verbal expression to this problem by saying that the series of episodes had made "standing alone" (remaining unmarried) painful to her.[9]

Freud's first full-length case study involved a grown woman, Dora, who had a number of conversion symptoms—difficulty in breathing, a nervous cough, and loss of voice—which Freud interpreted as having a common oral significance. The original traumatic experiences occurred when at the age of fourteen she became embroiled in a complex web of extramarital affairs between her parents and another couple. On one occasion the husband of the woman with whom her father was having an affair declared his love for Dora and kissed her on the lips. Freud speculated that she probably felt his erect penis pressing against her body at the same time. The symptoms that emerged in adulthood were the combined result of direct oral sexual excitation from the kiss and an upward displacement of the sexual excitation from her clitoris to her oral cavity. The transformation of her initial feelings of pleasure into disgust was possible because of the close anatomical proximity of the genitals and the anus. Excrement originally triggers disgust, and later on the genitals may act as a reminder of the excremental function. The oral focus of her symptoms was further determined by her relationship with her father. Freud rather boldly speculated that her nervous cough and sore throat were symbolic representations of her desire to perform fellatio on her father. She knew that he was impotent and had very likely heard that oral sexual intercourse was the only kind possible with an impotent man. Accordingly the tickling in her throat was a disguised hallucinatory fulfillment of her desire to have sexual relations with her father.

Since her choice of the oral cavity required some elaborate transformations of the original traumatic sexual experiences—displacement upward, disguised fulfillment of her desire for sexual relations with her father, and the transformation of pleasure into disgust—Freud added another process to help explain conversion hysteria: somatic compliance. Under stress from painful memories the mind generally seeks some psychic outlet, either normally in painful conscious memories or pathologically in the form of various psychic symptoms such as phobias or obsessions. But if there is some original bodily disturbance or fixation at an erogenous zone, a somatic conversion symptom may develop at that zone. In Dora's case Freud concluded that the mouth had become highly erotogenized when as a child she persisted in sucking her thumb. "An intense activity of this erotogenic zone at an early age thus determines the subsequent presence of a somatic compliance on the part of the tract of mucous membrane which begins at the lips." [10] The oral symptoms emerged in three stages. In infancy her fixation at the oral stage set up an organic basis for later symptomatology. At the age of fourteen she experienced traumatic excitation of her lips and, by displacement, of her entire oral zone. Twenty-five years later under conditions of psychic stress her oral cavity provided the organic setting and the somatic compliance for the oral symptoms.

Freud's subsequent full-length case histories concerned symptoms not primarily of a hysterical nature, but each drew from his general theory of the interrelation of mind and body. His interpretation of a phobic boy, Little Hans, in 1905, focused on the child's fear that if he were to go outside a horse would bite off his penis. In 1909 Freud published an account of an obsessional neurotic who was paralyzed by strongly ambivalent feelings about his father and a lady he admired. One of the precipitating causes of his obsession was his overhearing an account of a torture in which rats bore into a man's anus. He became obsessed with the idea of this torture being performed on both the lady and his father. His negative feelings were translated into fantasies in the somatic sphere. In that same case history Freud discussed the hypersensitivity to smell he believed to be particularly frequent among hysterical and obsessional neurotics. In 1910 he argued that visual disturbances might have a psychic origin. They might result from either defensive reactions against having observed something painful or punishment for indulging in voyeurism. "You wanted to misuse your visual organ to gratify your sinful lust. Therefore, it serves you right that you cannot see at all." [11] Freud's interpretation of the paranoid delusions of Dr. Schreber involved an elaborate set of delusions about the way his body was being transformed. (For a discussion of those symptoms see Chapter 10.) Though these were not conversion symptoms, they nevertheless illustrated how psychic conflict could be manifested in alterations of body image involving a central delusion that he was being emasculated and transformed into a woman. In a popular summary of psychoanalytic theory of 1917 Freud discussed bodily manifestations of ideas or impulses. One example he originally had written as a hypothetical elucidation of somatic representation of psychic conflict—the idea of marriage vows being represented by breaking of an arm—was confirmed in reality by a subsequent newspaper report that he ran across which illustrated this speculation exactly. A man accused of illicit relations with another woman confessed to the judge: "Today I feel compelled to make a full confession to the Court, for I have broken my left arm and this seems to me to be a divine punishment for my wrong-doing." [12] The broken arm was not a neurosis but clearly illustrated how mental conflict can register in the somatic sphere.

The various conversion symptoms Freud interpreted illustrated how parts of the body can become the location of manifestations of psychic stress produced by repressed traumatic material. When the neurosis is expressed in the bodily sphere, there is a relation between the precipitating cause and the resultant symptom. Freud once described that relation as a "mysterious leap" [13] from the mind to the body, and although he labored diligently to unravel the mystery, he never was able to trace the psychic or physiological pathways from mental activity to bodily symptoms. These remained a mystery to him.

CORPOREAL DETERMINANTS OF DREAM SYMBOLS

Freud explored another area of the mind-body problem with his interpretation of the psychic representation of bodily structures and functions. This work on symbolism matured in the course of analyzing his own and his patients' dreams in preparation for the publication of his classic *The Interpretation of Dreams* in 1900. To explain how certain elements of the dream are constructed out of latent material from the dreamer's childhood, he sought to discover the meaning of symbols in dreams and symptoms. He came to believe that these symbols recurred in all ages and places as part of a universal "unconscious ideation," which was particularly explicit among "primitive" people and which was found in folklore, legends, idioms, proverbial wisdom, and current jokes of all peoples.[14]

An early indication of his interest in dream symbols is a speculation he shared with his friend Wilhelm Fliess in a letter of 1897 in which Freud ventured to explain why witches fly in dreams: "Their broomstick is apparently the great Lord Penis." [15] This early interpretation of a phallic symbol was soon integrated into an extensive catalog of sexual symbols which Freud used to unravel the complex process by which his patients' dreams and symptoms were formed.[16]

The symbols Freud found most frequently appearing in dreams represented members of the dreamer's family or parts of his body. They included "the human body as a whole, parents, children, brothers and sisters, birth, death, nakedness—and something else besides." Parents appear as king and queen, and little animals or vermin may represent unwanted children or pesty brothers and sisters. Water symbolizes birth, traveling symbolizes death, and clothing and uniforms stand for nakedness. That "something else besides," he went on to say, was sexuality. "The very great majority of symbols in dreams are sexual symbols." [17] The number three may represent the entire external male genitalia, which has three parts. More commonly, male sexuality is represented by phallic symbols, which resemble the structure or the function of the penis. Structurally analogous phallic symbols include "all elongated objects, such as sticks, tree-trunks and umbrellas . . . as well as all long, sharp weapons, such as knives, daggers and pikes." [18] These latter reflect the penetrating function of the erect penis. Objects from which water flows, such as water-taps or springs, represent the urinary or ejaculatory function of the penis. The special magical property of the penis in being able to defy gravity provides another basis for symbolic representation for balloons, for birds, and blimps. Less obvious phallic symbols such as snakes, fish, and hats were confirmed by repeated observation of their symbolic meaning in dreams of numerous patients. The female genitals are represented by a similar set of objects which resemble their structure or function or which enclose or act as receptacles, such as caves, jars, and boxes. Jewelry boxes are particularly well suited to represent the female genitals because in German *"Schmuck"* may mean jewelry or penis, and *"Schmuck-kästchen"* has an obvious double meaning. Common symbols of the uterus are cupboards, stoves, or rooms; while the opening of the female sex organs may be symbolized by gates or locks. Keys are invested with phallic significance by virtue of their unlocking function. Somewhat less obvious symbols of the female genitalia include snails, churches, and the mouth. Pubic hair is symbolized by woods and bushes, and landscapes covered with complicated arrangements of vegetation and bridges may represent the anatomical topography of the female genitals. Castration may be symbolized by a dream of a tooth extraction or pulling a branch from a tree; masturbation, by knitting or playing a musical instrument; and the act of intercourse, by a dream in which the dreamer is dancing or climbing, particularly climbing a staircase. Freud explained this latter symbol by describing the likeness of climbing a staircase to intercourse: "We come to the top in a series of rhythmical movements and with increasing breathlessness and then, with a few rapid leaps, we can get to the bottom again." [19]

An account of the bodily determinants of *déjà vu* experiences which Freud added to the 1909 edition of *Interpretation of Dreams* was particularly daring, considering that Freud proffered his theory before a medical community that tended to dismiss his psychoanalytic theories precisely because they leaned so heavily on sexual determinants of human behavior. At times the dreamer has a strong feeling that he has been in the particular setting of the dream at some time before. "These places," Freud speculated, "are invariably the genitals of the dreamer's mother; there is indeed no other place about which one can assert with such conviction that one has been there once before." [20] This interpretation was scandalous. To suggest to the Viennese medical community of 1909 that a grown man places a dream in a familiar setting in order to re-experience his passage through the birth canal required a deep conviction of the validity of his theory of the biological determination of symbol formation. His dream theory offered the European scientific community a startling extension of the current understanding of mind-body interaction by demonstrating that the sleeping mind is preoccupied with many bodily processes—nutrition, digestion, excretion, urination, birth and death, as well as with sexual activity. The dream theory illustrated the general hypothesis that the human body determines a large portion of mental activity, and it underlined Freud's general conviction that sexual life plays an enormous role in human affairs.

BODILY DETERMINANTS OF CHARACTER

The same kind of thinking that went into Freud's theory of the bodily determinants of dream symbols also provided the foundation for his theory of the formation of character. Freud had explored mind-body interaction in the dreams and symptoms of his patients for many years before he attempted to work out a psychoanalytic developmental psychology. By 1905 he ventured to outline the course of normal development in *Three Essays on the Theory of Sexuality*, in which he argued that personality was profoundly influenced by the development of the erogenous zones of the body, which in turn determined specific character traits.

The first theoretical problem Freud encountered was the expansion of the going conception of sexuality beyond adult genital activity leading directly to orgasm. He stated that "it was necessary to enlarge the unduly restricted concept of sexuality, an enlargement that was justified by reference to the extensions of sexuality occurring in the so-called perversions and to the behavior of children." [21] Perversions illustrated that the sexual impulse can be modified in a number of ways to deviate from adult heterosexual activity by changing the sexual object (for the man) from the vagina to the mouth or anus, or even to regions of the body that are not orifices, such as breasts or feet. Clinical investigations showed that some men achieved orgasm without any interest in, or contact with, the vagina. (His account of perversions, by the way, is worked out exclusively from the man's point of view.) The sexual impulse may be further modified by altering the aim to involve such prefatory acts as looking and touching, or sadism and masochism. The impulse for perversion, he argued, is latent in everyone, and therefore the disposition for it must be created during normal development. He then postulated the shocking theory that the germ of perversion is present in all children. Child sexuality is "perverted" because it is focused on certain bodily zones other than the genitals and does not lead to intercourse. Furthermore it is auto-erotic and does not culminate in an orgasm. It would be a mistake to conclude from this analysis that children are perverts, but if their sexuality persists into adulthood and does not make the necessary transformation to produce adult heterosexuality, the result will be adult perversion.

Freud's theory of child sexuality has several parts. He argued that the sexual impulse is made of components which dominate child sexuality during various stages of development and which eventually become coordinated in the adult sexual function. These components are associated with certain erotogenic zones. During the first two stages, both boys and girls share the focus of their sexual activity around the lips and mouth, and then in the anal region. In the third stage, anatomical differences lead to gender roles when attention turns to the pleasurable sensations in the penis and the clitoris. The components also emerge in a specific temporal sequence. Each of these erotogenic zones dominates the sexual development of a particular stage, and they retain a subordinate influence on successive stages as each gives way to the dominance of the next zone. Thus, the components of the sexual impulse emerge additively and sequentially as they give way to increasingly complicated combinations of sexual feelings.

Freud's theory of child sexuality did more than merely describe the development of the sexual processes necessary to prepare the adult for the procreative act; it ventured to explain how certain character traits develop at each stage, and for that reason his description of this developmental process as "psychosexual" is appropriate. As he considered each stage he speculated about its characterological significance. This kind of speculation most clearly establishes his contribution, as I have viewed it in this chapter, as a discovery of the meaning of the body. Each erogenous zone triggers the development of character traits that *resemble* the function of that zone, and for that reason each zone has a unique meaning.

The oral phase is determined by the sucking impulse, and during it the child experiences the prototype of all later oral satisfaction from the pleasure of nursing and the feeling of satiety that comes from it. When the child learns to suck his thumb he establishes a certain measure of control over his erotic pleasure and lays the foundation for later character traits involving independence and autonomy. Adult pleasure from kissing, smoking, and drinking derive from this earliest form of infantile erotism. If the infantile oral experience is troubled, disturbances of the digestive process in adulthood such as disgust at the sight of food or hysterical vomiting may result. The ability to respond to the world is forged during this prototypical experience of absorbing nutritional material from the environment. Later psychoanalytic elaborations of the characterological significance of the oral stage would argue that the ability to love, to trust, and to learn is established during the oral phase. These character traits are meaningfully related to the biological nature of the oral experience: they involve the incorporation of foreign material, and the pleasurable experiences associated with it, into the mind or body.

The anal phase that follows involves a relation between the biological processes of retention and elimination of feces, and character traits that develop when the anal region comes to dominate the child's sex life. In an article of 1908 he explicitly related "Character and Anal Erotism." With calm assurance and appropriate caution he modestly began with a clinical report. "Among those whom we try to help by our psychoanalytic efforts we often come across a type of person who is marked by the possession of a certain set of character-traits, while at the same time our attention is drawn to the behavior in his childhood of one of his bodily functions and the organ concerned with it. . . . The people I am about to describe are noteworthy for a regular combination of the three following characteristics. They are especially *orderly*, *parsimonious,* and *obstinate.*" What element, Freud asked himself, did they have in common? "As infants, they seem to have belonged to the class who refuse to empty their bowels." [22] From this observation he concluded that these three character traits are most pronounced in people who derive unusually great pleasure from defecation or from withholding feces in childhood. And from his interpretation of the origin of these particular character-traits, he concluded, with regard to the whole range of possible character formations, that "permanent character-traits are either unchanged prolongations of the original instincts, or sublimations of those instincts, or reaction-formations against them." [23] This latter addition to his theory of anality applied to all of character formation and enabled him to derive any pronounced character trait, or its opposite, or some sublimated version of it, from a single determinative experience. Thus someone who was traumatized during his anal stage might become excessively stingy or excessively generous or develop some other modification of the personality that controls the exchange of valuable material between himself and others.

The connection between anatomy and character is particularly vivid in the anal stage. By learning to control his anal sphincter muscle the child begins to unconsciously acquire character traits that *resemble* that anatomical function; that is, he learns to hold back (obstinately) or relinquish (generously) his feces in the correct place and at the correct time, in conflict with his instinctual urge to release them wherever and whenever he wants but in accord with the moral and hygienic values communicated to him by his mother. Freud further speculated that feces has a similar significance for the child that determines its influence over generosity. It is dirty but valuable matter produced by the child's own labor. It is his first manufactured product. He values it as part of his own body and is reluctant to part with it, yet he does so in response to his mother's urgings. Feces is his first means of exchange and thereby becomes a symbolic first gift to his mother in return for her love. If this exchange takes place comfortably the child should not be overly concerned with his anal region and with excrement, but if it is fraught with anxiety, he may remain preoccupied with this bodily region. Such a fixation at the anal erotic stage may produce exaggerations of the character traits associated with this vital biological process. A fixated individual may manifest in adulthood any of a variety of character traits involving extreme stubbornness, stinginess, cleanliness, or their opposites—lack of will, prodigality, and dirtiness.

Over the years psychoanalysis has stirred a number of controversies. At its inception there was strong opposition to the idea that children have a sex life, one that Freud characterized as "polymorphous perverse." In the 1920s John Watson urged parents to restrict kissing and fondling of their children, in opposition to the psychoanalytic theory that children's emergent sexual instincts ought not to be excessively repressed. Recently a controversy has flared up over Freud's view of women. It is argued that his psychology was conceived from a man's point of view and tended to neglect the development of women, and where it did discuss women it characterized them as suffering from chronic and inescapable penis envy. To assess the validity of this critique, we must examine that part of psychoanalysis which commented on the sexual development in boys and girls at the conclusion of the anal stage, when the penis and clitoris take over as the foci of their respective sexual lives.

Sometime during their third or fourth year, children enter the phallic stage as masturbatory activity with the penis or clitoris comes to dominate sexual life. Freud's first discussion of this stage in the *Three Essays* said little about the characterological consequences of such masturbatory activity, except that the boy develops an "instinct for mastery" which will play an important role during his adult sexual life. He said nothing explicit about the girl's nascent genital sexuality except that both boys and girls have an innate sexual disposition like "an average uncultivated woman in whom the same polymorphous perverse disposition persists." He elaborated on this provocative piece of speculation by commenting that an "immense number" of women have an aptitude for prostitution and that the polymorphous perverse disposition must therefore be a "fundamental human characteristic." [24] By 1917 Freud expanded this account of the girl's clitoral sexual experience and argued that the girl's need to shift from clitoral to vaginal sexuality at puberty may account for the high number of sexually anesthetic women whose clitorises "stubbornly" refuse to relinquish their monopoly on genital sexuality. This explanation was further elaborated in a series of essays that Freud published beginning in the 1920s which attempted to supply a detailed account of the different course of sexual development in male and female children. The boy's sexual development is far simpler than the girl's for several reasons. He can retain his original libidinal attachment to his mother throughout his life without the complicated interruption that the girl experiences as a consequence of her rivalry with her mother for her father's love. The little girl also begins to resent her mother for endowing her with sex organs which she regards as inferior to a boy's. The development of the girl's sexuality is further complicated by a transfer of the focus of her genital sexuality from the clitoris to the vagina. Freud argued that "in her childhood . . . a girl's clitoris takes on the role of a penis entirely." [25] During the early phallic phase she attempts to achieve gratification like a boy by masturbating with her clitoris, but eventually she comes to realize that her organ is an inferior version of the penis and she begins to want a penis of her own. She experiences intense "penis envy," which dominates her emotional life. Her efforts to win a man and have a child by him are derivative of this early and futile desire for a penis. Another source of special difficulty for women is the wave of repression that inhibits their sexuality during puberty when the boy's emergent sexuality suffers no similar constriction.

Although Freud originally elaborated his developmental psychology from the man's point of view, he returned to the subject of female sexuality repeatedly and offered numerous reasons why woman's sexual development was the more complicated of the two. A number of the problems for women were created by unique historical and social circumstances such as rivalry with her mother and the repression that inhibits her adolescent sexuality. But several problems derived from the peculiar structure of her sexual anatomy. Freud made this general influence explicit in an article of 1925 entitled "Some Psychical Consequences of the Anatomical Distinction Between the Sexes," in which he outlined the general form of his argument from sexual anatomy to sexual character. "The difference between the sexual development of males and females at the stage we have been considering is an intelligible consequence of the anatomical distinction between their genitals and the psychical situation involved in it; it corresponds to the difference between a castration that has been carried out and one that has merely been threatened." [26] The boy experiences castration anxiety when he first views the girl's missing external genitalia and perhaps also if he is threatened with castration for masturbation, but the girl is plagued with the feeling that she has always been castrated and must remain so. Freud then offered some by now rather dated speculations about the character traits that form in women as a consequence of their acceptance of their castrated condition. "For women the level of what is ethically normal is different from what it is in men." They possess "less sense of justice than men, . . . [and] are more often influenced in their judgments by feelings of affection or hostility." [27] Freud made no logical connection between the acceptance of castration and ethical or emotional inferiority to men, but by arguing that women view the female sexual apparatus as inferior to a man's, the theoretical field was wide open for a negative assessment of female capabilities.

In "Female Sexuality" (1931), Freud presented his most complete summary of women's sexual role as derived from the structure of their sex organs. After surveying his earlier theory that a woman's sexuality is uniquely hindered by the necessity of having to renounce the clitoris for the vagina at puberty and her need to renounce her original love object, her mother, for her father, Freud turned to explore another consequence of woman's sexual anatomy —her bisexual disposition. Unlike the man, who has only one principal sexual zone, women have two, "the vagina, the female organ proper, and the clitoris, which is analogous to the male organ." [28] Accordingly the woman's sex life develops in two phases, the first of which is masculine, the second feminine. But because the clitoris continues to function in a phallic manner throughout life, women are more bisexual than men, and their sex life is more complex and more subject to dysfunction. A final source of anatomically determined character differences between the sexes is their respective response to each other's so disturbingly different sex organs. Men experience "castration anxiety" and women experience "penis envy." These contrasting emotional reactions, Freud concluded, give men a superior attitude. While men fear castration and often have a disparaging view of women because they are "castrated," women accept their "castration" as a given fact and acknowledge their sexual inferiority. Their response to this may involve rejection of sexuality altogether, clinging to their threatened masculinity with possible homosexual object choices, or intense love for their fathers and immersion in the feminine form of the Oedipus complex. Any of these alternatives further explains the more precarious process of sexual development that women must undergo.

Freud believed that female sexual anatomy helps explain women's higher incidence of sexual disorders. I would suggest that this interpretation involves some distortion of the given data and a blindness to some of the obvious advantages of female sexual life. From his supposition that women have two erogenous genital zones and men have only one, Freud concluded that men are less prone to disturbance in the course of development because of the relative simplicity of their sexual development. Why not reason by simple addition and conclude that two is better than one? He said that women's need to renounce their clitoris leads to conflict about the true center of their sex life. But since the clitoris continues to function, why not conclude that women have an advantage because they have the more variable sexual apparatus? Why is female sexual complexity necessarily a detriment? When elaborating on the superiority of the external male genitalia, Freud argued that its imposing complexity is viewed by women with awe. Why is that kind of complexity an advantage, while woman's complex sexual development is not? Why not view her more bisexual disposition as an advantage, offering her a wider range of sexual feelings and a greater capacity to make do with the necessary limitations of monogamous sexual relations? Why not mention woman's obvious sexual superiority over men by virtue of her ability to have intercourse when not completely aroused and to have a series of orgasms in rapid succession? By contrast, the man is without question biologically inferior. When he is not aroused he is totally incapable of the sex act, while women can at least "fake it." This is not an ideal situation, but it at least enables a woman to perform and get through emotionally difficult periods without causing a traumatic situation for herself, while the impotent man is, in his own eyes and in his partner's, a complete failure, with his failure quite visibly on display. Woman's greater physical capacity for sex, her capacity to have multiple orgasms, and the increase in her sexual appetite in her late twenties all contrast with a more limited physical capacity in man. He can have sex less frequently than a woman, he is generally capable of single orgasms only, and his sexual powers begin to decline when a woman's are still on the rise.

The limitations in Freud's understanding of the biological inferiority of male sexuality, however, does not justify the kind of attacks on his theory of sexuality that have been made in recent years. His pioneer work into the manifold ways sexual anatomy determines character offered a mode of explanation about the origin of character that was impressively original and has proved to be enormously fruitful. Even though much of his gender psychology degrades female character, he did much to assist the women of his age to understand and deal with their very real problems. At a time when few physicians ventured to even admit the existence of sexual instincts in normal women, Freud insisted that their sex life must be understood and taken seriously. His early patients were women suffering from problems relating to their sex life, and it is therefore understandable that Freud would tend to emphasize the sexual origins of all neuroses and would tend to view female sexuality as particularly prone to creating mental problems. Breaking down the rigid views of manliness and femininity as mutually exclusive and polar opposite character structures, which they were so narrowly conceived to be in the late nineteenth century, Freud argued for the basic bisexual disposition of both sexes and implied that the healthy personality was a mixture of the masculine and the feminine. His by-now notorious theory of penis envy no doubt exaggerated this ingredient of the female mind, but, if his clinical reports are truthful, it was nevertheless a real emotion experienced by many of his patients. Freud did overlook the social origins of much of woman's envy of man, but his insistence on the biological origin of woman's envy as caused by her perception of her own and a man's external genitalia was a most imaginative and original contribution to the history of ideas.

CIVILIZATION AND THE BODY

Freud's thought on history focused on the fate of man's instinctual life. The conclusion he consistently drew was that the progress of civilization demanded ever more instinctual renunciation and left human physical existence ever more repressed and "uncomfortable." In a letter to Fliess of 1897 he wrote that progress demanded renunciation of the incestuous impulses that dominated primitive societies, and then in 1905 he commented on the "inverse relation holding betwen civilization and the free development of sexuality." [29] In an article of 1908 he viewed the problem as a cause of the rise in the number of cases of mental illness. We can control and sublimate the sexual instinct to a degree, he said, but "for most people there is a limit beyond which their constitution cannot comply with the demands of civilization." [30] The body can take just so much repression. Though his first patients were largely women suffering from various negative responses to this excessive imposition of sexual repression, he concluded with a telling comment about "how rarely normal potency can be found in men" living under the sway of civilized sexual morality.

In an article of 1912 Freud explored more fully the fate of male sexual potency in the grip of civilized sexual morality. There is an essential sensuous component of erotic life that tends to atrophy in a culture that exaggerates the affectionate components. A large number of men complained to Freud that they were able to achieve full potency only by degrading their sex object, and they had great difficulty with their well-bred wives. As a consequence, he wrote, "psychical impotence is much more widespread than is generally supposed." From this fact he concluded that "the behavior in love of men in the civilized world today bears the stamp altogether of psychical impotence." [31] There may be no way at all to reconcile the sexual instinct with the demands of civilization.

Freud alluded to the problem a number of times in subsequent works and finally developed it into a full-length study, *Civilization and Its Discontents* (1930). His generally pessimistic view of the destructive effect of civilization on sex was unchanged; in this work, however, he was able to expand his account of the many ways instinctual life is repressed, and alongside of the erotic component of the sexual instinct which had been his focus in earlier studies, he now added an account of the fate of the aggressive instinct as well. Basic functions in a civilized society such as personal cleanliness, orderliness, and beauty require instinctual renunciation. The rule of law in a society forces man to renounce many of his aggressive impulses toward others, and the demands of monogamy in a civilized society require enormous sacrifices of instinctual sexual urges.

In his essay of 1908 Freud was preoccupied with the effect of this renunciation on the genesis of individual cases of mental illness, but by 1930 he was deeply concerned that the repression of instinctuality would generate a massive load of guilt which would someday be acted upon in the form of destructive impulses directed against civilization itself. The response of the sons to the authority of the fathers, which Freud had discussed in *Totem and Taboo* (1912-13), was to rise up and slay the primal father, and so perhaps mankind might rise up and destroy the source of its own guilt—the civilized morality that imposed it. In 1915 he had interpreted the outbreak of World War I as a result of excessive instinctual repression imposed on a generation of Europeans living "beyond their moral means," under a load of repression that had become too much to bear. In that war the long-suppressed murderous impulses lost the stigma of taboo and became the guiding principles for the action of states. European authority had been fatally crippled by its own rebellious sons. Freud concluded *Civilization and Its Discontents* with a hope that the forces of the erotic instinct would prevail over the destructive instinct and work to preserve civilization. But that final positive hope was based on a contradiction of the argument which had preceded it, because Eros itself, Freud had just argued, is subject to ever more repression, and it too is enfeebled as civilized morality becomes dominant. The theoretical tangle that concluded *Civilization and Its Discontents* was sufficiently engaging to lead a number of thinkers to attempt some resolution of it. Herbert Marcuse in *Eros and Civilization* (1955) acknowledged the tension between the forces of Eros and the demands of civilization. He suggested that a social revolution might be able to restructure society sufficiently to allow for a partial discharge of the instincts through some kind of "non-repressive sublimation" to avoid the destructive consequences that follow from excessive repression. Norman O. Brown explored the dilemma in *Life Against Death* (1959) and offered the hope of a "resurrection of the body" to rekindle repressed sexuality.

The simple argument that Marcuse and Brown found so challenging—that severe moral demands generate guilt, and severe guilt leads to self-destruction—was first worked out in similar terms by Nietzsche in the 1880s, but Freud's presentation of it captured the interest of so many because it was part of the imposing intellectual edifice of psychoanalysis that he had developed to deal with the entirety of human existence. I have focused on a major part of that system that belongs in the history I am tracing—changing views of the body. No single thinker did so much to influence the way modern man conceives of and experiences his corporeal existence. Charcot had drawn a chart showing the various hysterogenic zones, but it was Freud who first described the exchanges, displacements, and substitutions of them that were possible. To understand the way psychic stress can create conversion symptoms, he first had to decipher the meaning of dreams and symptoms, and in the process the workings of the unconscious mind were revealed to him. The body itself, he found, is a source of ideas and character traits as well as direct somatic stimuli, and it operates on mental life in a meaningful way. There is a silent language between the body and mind that Freud translated into easily comprehensible terms. To the dismay of his readers he insisted that the possession of an anus and the experience of the anal function generated character traits which, until that time, had seemed totally independent of "anal erotism." The audacity of this extraordinary argument was matched by a clarity of thought, a precision in expression, and a wealth of clinical experience that gave it a great deal of attention among European thinkers, and the theory of anality was just a part of a larger system of thought which may deservedly be called a discovery of the meaning of the body.

Chapter 15

Eros in Barbed Wire

The generation preceding the outbreak of World War I witnessed a sweeping transformation of the ideas and values that had held their world together for centuries. The revolution in artistic conventions, the revision of the theoretical foundations of the social and the physical sciences, and the erosion of the social and political structure intensified the impact of the revolution in sexual morality that was under way and produced an atmosphere of general cultural collapse. One historian has characterized it as an era of "dissolving certainties," although during the period Europe produced some of its most impressive cultural achievements.[1] Modern art emerged from the Cubist movement, and by 1911 painters were producing non-representational paintings. The rules of rhythm and harmony that had governed musical composition for centuries were set aside in the search for new combinations of sounds in the twelve-tone scale. Social scientists began to abandon the search for complete objectivity and acknowledge that certain irrational forces were operative in human affairs.[2] Just as the painters had given up trying to render what was "out there" and had begun to express their own subjective responses to the world, so did historians give up the search for objective knowledge about the past. The positivist historian Leopold von Ranke's injunction to describe the past "as it really was" gave way to a more subjective approach. Sociologists came to focus on the operation of imitation and irrational group pressures in social phenomena, and, following Freud, psychologists began to explore the workings of instinctual impulses and unconscious processes in mental life. Even that last bastion of certainty and order, the Newtonian physical universe, came to be viewed as a world of hypothetical entities, irregular movement, paradoxical phenomena, and theoretical uncertainty. The revolution in physics revealed that matter was largely empty space, not the solid stuff that common sense supposed. Particles moved through space in unpredictable patterns, and as they moved they shrank in size. Matter and energy were interchangeable, and mass increased when accelerated. Light behaved at times as a stream of particles and at other times as a wave, and the speed of light was constant and therefore violated the Newtonian law of the addition of velocities.

The enormous social changes that were slowly brewing in the nineteenth century accelerated in the prewar years and anticipated some of the major developments of the war. In the course of the war three monarchies collapsed and with them the heart of the aristocratic life style. Titles of nobility and the social privileges that were associated with them would never again have the same significance in European social life. Another casualty of the war was the sexual morality and the traditional power relations within the family. In this chapter I will focus on that transformation in sexual morality and the changing attitudes toward the body that emerged from the war experience.

Although historians of morality disagree in their evaluations of the moral revolution that anticipated the war, they do tend to agree that one took place. The core of that revolution was a substantial revision of male-female relations in the direction of female emancipation and its consequence—the emergence of more aggressive female sexuality. The sexual suppression of women throughout the nineteenth century had been linked with their exclusion from society and their seclusion in the home, and so their emergence from the kitchen and the nursery into professions long held to be the exclusive preserves of men brought about a concurrent emergence of their sexual life and a restructuring of power roles within the family. Some writers have speculated that the realignment of male-female relations contributed to the tensions that led to the outbreak of the war and intensified the suicidal frenzy that was so widespread during the first months of fighting.

Two years before the outbreak, an Italian sociologist, Emanuele Gallo, had published a book which dismissed the economic and social causes of war and argued that all wars were caused by sexual problems alone.[3] Magnus Hirschfeld disagreed with Gallo's insistence on sex as the only cause of war, but he did affirm that the impulses that led to war were intensified by propaganda with strong erotic undertones.[4]

A number of studies noted the importance of the revolution in morals leading to war. By 1918 it was clear that the old moral order had broken down, and the apparent collapse of the prewar morality seemed a most plausible explanation. Some believed that the changes were long overdue and brought relief from the excessive repression of the prewar period, though many others viewed the moral changes unfavorably as a major cause of the final explosion. In 1917 the German historian of morals Bruno Grabinski surveyed the numerous ways the moral breakdown had corrupted Germany. He faulted movements for the defense of homosexuality, free love, nudism, and feminism. In those prewar days "a moral syphilis" had spread over the nation. "The filth piled higher and higher until the stink rose to the heavens—no class, no age group was exempt."[5] An unrestrained "sexualism" was at the center of this progressive contamination of German moral fiber. The statistics on prostitution, venereal disease, and the sale of pornography in the prewar years showed that German morality was in mortal danger.

Another German critic, E. F. W. Eberhard, blamed Germany's moral and cultural crisis on the progress of a special "female sadism" which sought to destroy men. The figures of the vampire and the *femme fatale* appeared frequently in the art and literature of the period, as Eberhard and numerous others observed.[6] The "divorce dramas" particularly alarmed the critics. The work of Ibsen, Strindberg, Wedekind, and Tolstoy implied that sexual relations were an inevitable and interminable battle and that marital fidelity and family happiness were hopeless dreams.

Some moralists viewed the war as a welcome relief from the steady progress of moral decay caused by the feminists. In an article of September 1915, Karl Scheffler characterized the war as a struggle between the manly Germans and the effeminate French and English. Toward the end of the war a German physician expressed the similar hope that the enforced continence and the consequent "reabsorption of semen by the blood" would be a "steel bath of nerves" for the soldiers in the field.[7]

Henry Miller commented on the sexual origins of the war in his unique speculative style. "When doubt and jealousy run amok it is because the body has been defeated, because the spirit languishes and the soul becomes unloosed. Then it is that the germs work their havoc and men no longer know whether they are devils or angels, nor whether women are to be shunned or worshipped, nor whether homosexuality is a vice or a blessing. Alternating between the most ferocious display of cruelty and the most supine acquiescence, we have conflicts, revolutions, holocausts—*over trifles, over nothing*. The last war, for example. The loss of sex polarity is part and parcel of the larger disintegration, the reflex of the soul's death, and coincident with the disappearance of great men, great deeds, great causes, great wars." [8]

Though evaluations of the moral revolution varied from one observer to the next, several general conclusions about the significance of the changes were drawn by its critics and supporters alike. They agreed that major moral changes were under way in the prewar years and that they had contributed to the outbreak of hostilities. Those who viewed this revolution favorably argued that the war helped complete the changes under way, in particular recognition of the rights of women to greater sexual freedom. Those who criticized the changes nevertheless believed, at least in the early months, that the war would improve the moral situation in Europe and bring about something like the "steel bath of nerves" that Scheffler had predicted. When they were disappointed in the latter years of the war they offered dire predictions of complete moral collapse.

Recalling his own experience fighting in the Spanish Civil War, George Orwell observed that soldiers develop an acute sense of body awareness. "You always, I notice, feel the same when you are under heavy fire—not so much afraid of being hit as afraid because you don't know *where* you will be hit. You are wondering all the while just where the bullet will nip you, and it gives your whole body a most unpleasant sensitiveness." [9] The soldiers exposed to shelling in World War I must have acquired a similar hypersensitivity. The physical discomfort of trench life, the horror of seeing comrades torn to pieces, the constant fear of being shot continually focused attention on their bodily existence. The moaning of the wounded further underlined the fragility of the body and its vulnerability to pain, and the smell of urine, excrement and gangrenous bodies filled the atmosphere with constant reminders of the organic nature of human existence. For the civilian population, scars and missing limbs recalled the agony that the crippled bodies had once endured during the slaughter. The final calculation that ten million had been killed provided a chilling conclusion to this chapter in the history of human destruction.

One German observer during the war commented on the moral significance of the sacrifice of men's bodies. "German men subject their bodies to unspeakable strain and fatigue as a sacrifice for the Fatherland, while German women and girls at home make a sinful show of their all too thinly covered bodies." In 1919 one German physician sanguinely reviewed the progress that medical sciences had made from experimentation on the wounded. Medical technology had developed artificial limbs and supporting apparatuses for the crippled; knowledge of hygiene, infection, and inoculation was advanced; various surgical techniques were perfected, in particular operations on the muscles in stumps.[10] The vast store of information gathered on venereal disease, abortion, and contraception further contributed to the progress of medicine. The lack of privacy influenced personal hygiene, body image, and sexual life. One historian wrote: "In the environment of the military hospitals, where all the secrets of the body had perforce to be exhibited without reserve, it is only understandable that everything pertaining to love should have been treated with a considerable measure of indulgence." [11] E. M. Remarque's classic war novel *All Quiet on the Western Front* included an episode in which a soldier had sexual intercourse with his wife in a military hospital in the presence of his neighbors in the surrounding beds. The English feminist Mary Agnes Hamilton explained how female sexual mores were affected by the repeated observation of suffering and mutilation. "The religious teaching that the body was the temple of the Holy Ghost could mean little or nothing to those who saw it mutilated and destroyed in millions by Christian nations engaged in war. Little wonder that the old ideals of chastity and self-control in sex were lost."[12] This questioning of the value of chastity was part of a larger revision of the sexual roles within marriage.

In the course of the war the institution of marriage lost some of the respect that had sustained it over the centuries. At the outbreak of the war couples rushed to get married after short courtships. Justices of the peace in England reported having to marry eager couples in groups of twenty to accommodate the demand. The rise in the number of hasty marriages during the early days of the war was matched by an equally high, and even more alarming, rise in the number of divorces sought during and after. The long separations of husbands and wives intensified the problems created by changing sexual mores. Already in 1916 the German feminist Helene Stöcker complained that the war had complicated male-female relations by fostering misunderstanding and lying. The enormously different experiences of the sexes tended to alienate husbands and wives as understanding gave way to suspicion and as men sought to restore their diminishing authority within the family.

Several studies focused on the effect of the war on male sexuality. The German physician P. Lissmann reported that prolonged sexual abstinence led to frequent nocturnal emissions, giddiness, and "strange sensations in the testicles."[13] He found that many soldiers on leave were impotent following their return from the front. Some began to lose control of their sexual organs during artillery attacks and experienced involuntary ejaculations. Lissmann speculated that this particular sexual aberration showed the link between sexual excitation and fear, so exaggerated by battle conditions. A number of men remained impotent as a result of their war experience.

The extreme physical discomfort of life at the front affected general attitudes toward the body. The verminous clothes, the filth, the drunkenness, and the need to defecate and urinate without privacy obliterated delicate sensibilities. The forced sexual abstinence led to a progressive deterioration of the nerves rather than the "steel bath of nerves" that had been predicted. Fond memories of tender sexual relations were soon replaced by obscenity and pornography. Magnus Hirschfeld concluded that the war stimulated sadism with strong sexual content. He wrote: "The many senseless acts of cruelty of the World War are inconceivable without sexual underpinnings." [14] Drawing from Freud's theory of the instinctual link between erotic and destructive impulses and their common repression in civilized society, Hirschfeld argued that the war allowed a resurgence of both impulses, a "return of the repressed." An Austrian physician estimated that ten per cent of the Hungarian cavalry in his sector practiced sodomy with the horses entrusted to them.[15] As the war dragged on, fantasies about women became more vulgar and more brutal. The lack of privacy and the intimate contact with only masculine sexuality stimulated pathological auto-erotism, homosexuality, and transvestism. Upon returning home, many wartime homosexuals found it impossible to return to heterosexual relations. Others were impotent, and a large group suffered from premature ejaculations which, according to several medical reports, lasted for many years.

The heterosexual life of the men at war was limited largely to prostitutes. In addition to the usual causes of the rise of prostitution during war, the First World War presented two unique factors which made it such an enormous problem. Never before had such large numbers of the male population been in uniform, and after some initial maneuvers the armies of the two sides settled into a stable trench warfare which allowed prostitutes to set up bordellos. Some mobile bordellos followed the troops about, but most of the traffic was handled in special houses, established not far from the front lines. The military authorities encouraged the use of such facilities to control the spread of venereal disease, which soon became a major military problem. The prostitutes in these special houses were inspected regularly, and in June 1915 the German authorities ordered that any prostitute who knew that she was infected with venereal disease and had intercourse with a soldier would be given from two months to one year of prison.

The mechanization of life had indeed penetrated the sexual sphere. Hans Henel's war novel, *Eros in Barbed Wire* (1931), described a typical experience at these special houses. "Every soldier had to show his member to a sanitation officer; examination was made for morbid phenomena and treatment of protargol and vaseline administered. Thus fortified, the soldier entered the brothel. When he returned, he was obliged to urinate in the presence of the sanitation officer in addition to receiving another protargol injection." [16] Another historian of prostitution recalled that the men standing in line outside the house were instructed as they got near the door: "If the pipe is not stiff, stroke it for a while!" [17] The sex life of the soldiers at the front was thus socially controlled by military authorities, distributed mechanically like food in a mess line, and rushed into the space of a few minutes' time. Most prostitutes were able to handle twenty to thirty men in a working day, and some are reported having had fifty or sixty. Respect for the female body as well as for their own suffered a steady decline during the years of front-line fighting, and for those soldiers who did not experience impotence or premature ejaculation upon returning home, resumption of the more delicate prewar sensibilities was impossible.

There were, of course, special houses for the officers, which were usually marked with a blue lantern, while enlisted men visited houses with the traditional red lantern. Such separations stemmed from the practice of denying lower class men access to upper class women. If the war was being fought even in part to defend the European upper class man from threats to his sexual and social hegemony, then even the unfortunate sex objects who worked in the bordellos had to be protected from contamination by enlisted men. The women who serviced the officers were therefore endowed with this special status, because contamination of one's body from contact with the same prostitute constituted a serious threat to the maintenance of class differentiation. It is unfortunate that we have no records of the conversations between officers and their companions as they joked about the goings on in the enlisted men's bordellos, but one visual report has survived from a man in the sanitation service in a military brothel in Mitau. "In the officers' brothels," he reported, "mad scenes sometimes took place. What respect could we have for our superiors when we saw officers' faces being slapped by or spat on by brothel girls?" The account went on to describe naked brothel girls riding on the backs of uniformed officers on all fours. It concluded by questioning how men could admire, let alone follow into battle, officers who behaved in such a manner.[18]

In addition to the psychological degradation of sex, war produced a more tangible kind of sexual problem—venereal disease. In spite of the military's attempts to control it by instructing the men of its dangers, requiring them to use preventive measures, and threatening prostitutes with prison for infection of soldiers, the disease quickly grew to epidemic proportions and began to pose a serious military problem by reducing the effective fighting force of military units. Although at the time it appeared to be out of control, the educational and preventive measures introduced during the war proved to have lasting value in the postwar years. The magnitude of the problem and the exigencies of a wartime situation forced Europeans to set aside some of the restrictions that had prevented open discussion of venereal disease before the war. One historian has succinctly concluded that the war "spread promiscuity upwards [socially] and birth-control downwards." [19]

The war had as profound an effect on women's sexual lives as it had on men's. In the first weeks the women shared the exaltation of the men, and in many instances that feeling was manifested in the sexual sphere. The war tapped new sources of female sexuality and offered new outlets for its expression. The general atmosphere of danger and emergency temporarily rolled back the forces of repression, and women experienced heightened eroticism. Women, aroused by masculine displays of strength, were particularly excited by the marching men and the ceremony. This immediate response tended to subside, but a number of important positive consequences were registered in the later stages. The new professions that opened up for women because of the shortage in manpower also tended to open new sex roles. Nurses in particular developed a fuller knowledge of men's bodies. Not every woman derived conscious pleasure from gazing upon wounded men, but I do not think it is too speculative to suggest that the nation as a whole lived in a more sensuous involvement with the bodies of the men who were exposing themselves to dangers on the field of battle. The nurse tending the wounded soldier became a symbol of woman's contribution to the war effort, and that symbolic role involved satisfaction of long-repressed voyeuristic impulses.

Clothing had remained extremely modest until the war years, when, for the first time in a century, high fashion began to reveal women's ankles and legs. The same factors that worked to ease the restrictions on sexual morality allowed women to display sexually attractive parts of their bodies and forced them to compete aggressively for the few eligible men at home. The scarcity of materials may have also added to the incentive for shortening women's dresses and saving cloth, but the major factor was the changing sexual morality of the times. Hemlines rose off the floor and necklines plunged ever lower. Provocative colored underwear was introduced to replace the traditional white linen, and women rejected the idea that only prostitutes wore cosmetics and began to appear in lipstick, rouge, and eye makeup. Women's clothes also began to imitate men's. After the war male uniforms and male civilian styles continued to influence women's fashions. The changing sexual mores were visibly on display in the fashion of the 1920s.

Female sexual aggressiveness was able to surface as the chaotic circumstances of chance encounters made elaborate and extended courtships impractical. Women were emboldened to express sexual aggressiveness when their husbands and lovers returned home, and, after a long separation, the men were less likely to recall or impose the prewar restrictions on the expression of a woman's love.

Some sexual responses of women to the war bordered on the pathological. A number of German psychiatrists reported that some women developed a special attraction to prisoners of war. Wilhelm Stekel speculated that they found these prisoners particularly lovable because they shared a common enemy—German men. A number of women are reported to have become excited by men's uniforms, and a special diagnostic category—*Uniformfetischismus*—made its way into German psychiatric textbooks. Another woman aroused herself with the braids on her husband's uniform. These sexual responses did not occur in large numbers, but the particular forms that they took reveal some of the pressures that were operative on the population as a whole.

While most women experienced moderate elation during the first weeks of the war, some went sexually berserk. Another diagnostic category—war nymphomania—was introduced to describe them. One woman masturbated to the sound of marching troops. Some were particularly excited by reports of cruelty or the actual witnessing of it. Hirschfeld recounted the case of one woman who became sexually aroused when viewing her husband's battle wounds. Gynecological reports reveal that large numbers of women experienced a wartime cessation of their menstrual period. A

German gynecologist suggested that these cases of "war amenorrhea," as they were called, were caused by a number of wartime experiences—malnutrition, sexual abstinence, overwork, and particularly shock from learning of the death of a husband at war.[20]

While the men were indulging themselves in brothels, women began to explore the previously forbidden experimental terrain of pre-marital and extra-marital affairs. Like the men, some women became promiscuous. They shared with the men the subsequent guilt, but for the women there was the added danger of pregnancy and the prospect of a living reminder of their wartime adventures. Since abortion was illegal and a source of shame, the statistics on it in this period are scant and unreliable, but all observers noted a sharp rise in the number of women seeking some solution in this manner. Even more shocking were the infanticides committed by women too frightened to abort their child but equally terrified at the thought of having to confront their returning husbands with an illegitimate child.

A number of the criminal acts of the war years were triggered by the sexual rivalries and jealousies created by the long months of separation. Many husbands on leave, and particularly on unexpected leaves, returned to find their wives in adulterous liaisons. The returning husband who murdered his wife or her lover was usually acquitted on the grounds of "justifiable homicide." Some of the jealous men committed suicide, while some few tortured their wives. One German soldier forced his wife to remove her clothes and then sit on a heated stove and singe her genitals. This method of reprisal was imitated by several other men. Some men intentionally infected their wives with syphilis. One woman attempted to prevent the return of her husband by sending him a packet of cookies made with rat poison.[21] Unmarried women were frequently given prison sentences of from two to six months for having had intercourse with enemy soldiers or prisoners of war. A few cases are reported of women having castrated the bodies of dead soldiers.[22]

These ghoulish acts constituted but a small part of the large-scale sadism tacitly sanctioned by the war-crazed population as a whole and quite consciously carried out by the military leaders. The war propaganda controlled by the Allies included stories that German soldiers cut off the breasts of Belgian women and bayonetted babies. Acts of cannibalism, disemboweling, and rape were frequently reported in the press of all belligerents, though they were doubtless embellished. Russian troops apparently did rape a number of Galician women, and the Turkish troops certainly did slaughter two hundred Armenian women one night while German soldiers stationed nearby did nothing. These spectacular atrocities aroused the moral indignation of the European population at large, but their moral blustering did nothing to avert the general slaughter in the trenches. The European moral conscience was considerably dulled by the enormous destruction of human bodies. Commodities in abundant supply quickly lose their value, and human bodies—dead or alive—went cheap during the Great War.

Those men lucky enough to return in good health to receptive and loving wives had to learn to adjust to enormously different morals, aesthetic sensibilities, and sexual relations. The sexual supremacy of European men was on the line in the war. Already, in 1915, the feminist German writer Grete Meisel-Hess protested a bit too much with her claim that a good woman can live celibate without duress.[23] One journal on venereal disease of that time argued in the same vein that women had only ten per cent of the libido of men.[24] But such speculations were of little comfort in the face of the substantial changes in the direction of fuller and more active sexuality for women that were accelerated by the war. German men, facing military defeat, were particularly sensitive to the new attitudes of women. The literature on manliness and sexual morality that appeared in Germany in the latter years of the war reveals that the German masculine ego, traditionally quite intimately linked with military excellence, was at an all-time low in the castrating atmosphere of military defeat.

Emil Gustav Paulk's guide to manliness, *Die Manneslehre* (1918), was a catalog of male sexual fears. There is, he argued, an instinctual sadistic component of the female mind that manifests itself in nature by preying on males and killing them after they have performed their sexual function. In women this instinct surfaces as psychological vampirism, and it is particularly evident among the feminists. Women formerly received a man as a city receives a conqueror. She opened her body to him fully, in complete surrender, but with the current use of contraceptives men are robbed of their true function and become impotent. Paulk conceded that "women are more sexual than men," and for that reason men must subdue them. Their effort to re-assert control over female sexuality is hindered by Jewish men, who are extremely sensual and constitute a special threat to fair German men. For the Jew the most vulgar sexual acts are routine. "The limitations of shame and inhibition which restrict the Aryan are unknown to the Jew." [25] The Jew is a natural seducer, and he is able to make German women forget their modesty and engage in the most intimate sexual practices. German men are "poor fools." They respect women and thereby lose out to the exotic, the brutish, and the shameless men of the inferior races. The aggressiveness of modern women strips men of their power and renders them impotent, and that impotence further drives European women to seek out the other more potent races. To break out of this dilemma, Paulk suggested that Europeans learn from the Orientals how to prolong the act of intercourse and wait for women to have *"Endeffekt."* Paulk concluded with the interesting historical speculation that "this need to wait for the woman is perhaps new at this time." [26] Maybe it was. Since the 1880s medical and feminist literature had been exploring the sexual needs of women and demanding their satisfaction. Some men went off to war, I have argued, at least in part to work off the frustration of unsatisfactory sexual relations (as men do in all wars). During their absence the women at home were forced to evaluate carefully their sexual and emotional requirements. The sexual revolution that took place during the war made a return to the prewar sexual conventions impossible. When European men and women began to confront one another once again under peacetime conditions, they were forced to deal with a new consciousness of the sexual needs of women. Some began to rethink their old values and abandon them, while others held tenaciously to the sexual mores of the life style that was dying with the end of the war.

E. Heinrich Kisch made a heroic defense of the traditional view of male sexual superiority in a pathetic two-volume treatise titled *The Sexual Infidelity of Women*, published in 1918—the year of Germany's defeat. Following a proud restatement of the logic of the double standard, he rehashed some conventional ideas about women's being naturally modest and timid and drawn to men possessing "elemental strength and a strong will." Invoking the rhetoric of the military machine that was crumbling all about him,

Kisch appealed to men to hold on to their "Herculean manliness" and rule with a "strong fist" to appeal to women's true sexual fantasies. The powerful conqueror, the tyrannical master, the uncompromising strong man, he claimed, most fully satisfies the erotic needs of weak and dependent woman. But this ideal sexual arrangement of master and slave bound together within the traditional marriage is breaking down because of such recent cultural developments as erotic books, nudity in art, and the preoccupation with divorce in the theater. Crowded city life further stimulates female sexual instincts and undermines marital fidelity. Even chronic constipation in women can tend to heighten undesirable sexual excitations. The report of two physicians who estimated that forty per cent to fifty per cent of women have no sexual feelings he rejected as "too high and completely false." A woman's sex drive is much weaker than a man's and normally develops only in marriage. If this were not so, as Krafft-Ebing argued, "the world would become a brothel, and marriage and the family would be inconceivable." [27]

Kisch then listed several kinds of women who tend to break up marriages. The *over-educated* woman is often infertile and unhappy in marriage. The *childless* woman seeks divorce as the only way to re-marry and have children. Kisch acknowledged that eighty per cent of all childless marriages are caused by men, generally because of some previous or current infection by gonorrhea or syphilis. The *morally degenerate* woman never develops proper maternal instincts and is ill-suited to marriage. Many of these women are congenitally degenerate and manifest overt physical signs such as deformed ears, crooked teeth, or dark hair on the upper lip. Their sex life is out of control usually because of excessive masturbation during childhood and adolescence. Certain *overly congenial* women have highly developed sexual instincts and are easily tempted into infidelity. The *emancipated* woman is corrupted by feminist arguments into becoming dissatisfied with marriage. Her effort to achieve an "emancipation of the flesh" leads to her emancipation from marital fidelity. Some of Ibsen's heroines illustrate another kind—the *coquette*—who, like Ellida Wrangel in *Woman of the Sea*, is incapable of remaining true to her husband.

Kisch explored the numerous causes of this growing infidelity in women for over two hundred pages before mentioning the war and its impact on male-female relations. He looked to the war experience to improve marital relations—it would, he hoped, lead to an increased sense of marital fidelity and clarify the value of family and the state. He did not comment on the significant rise in the divorce rate during the war years, nor on the many changes in sexuality that it brought about that made the traditional relations impossible to restore.

His ideas offered a comprehensive picture of the kinds of insecurities that troubled European men in the wake of the war. Many men would not acknowledge for years to come that women were making bolder sexual demands, while others had begun to understand and accept the changes some time before. Ibsen had dramatized a power shift within a marriage in *A Doll's House* more than thirty years earlier, but the war was the major turning point in the evolution of sexual and marital relations. It intensified and accelerated the changes that were under way and presented them to millions in a sharply focused period of time—a period that transformed European life in so many ways. Lewis Mumford has given a colorful account of the recrudescence of sexuality following World War I. "When Mars comes home, Venus waits for him in bed . . . and to the degree that she has been neglected during the war she demands compensatory attention in times of peace. . . . And she leaves nothing undone to gain her end: she exposes her breasts, she takes off her undergarments, reveals her limbs, even *mons veneris* itself, to the passer-by. . . . The women's styles that prevailed in the Western World at the end of the last martial debauch match almost point for point those that became fashionable at the end of the Directory—down to the removal of the corset and the temporary abandonment of the petticoat." [28]

Along with revisions of the territorial boundaries and the balance of power among nations that were made after the war, sexual relations were also restructured. New anatomical territorial boundaries were established for purposes of sexual attraction. Women revealed more of their neck and bosom and legs than they had before the war. Several accounts noticed that both men and women walked more freely and carried themselves less formally after the war. New power relations were also established between the sexes. The older balance of power, weighted so much in favor of men, was shifted to accommodate the claims of women, and the traditional active-passive masculine and feminine roles collapsed toward the middle. As the theories of psychoanalysis began to be accepted after the war, new erogenous zones were explored and integrated into new acceptable patterns of love-making. Foreplay became more innovative, so that by 1926 Van de Velde's suggestion of using the genital kiss for purposes of sexual arousal was widely discussed. Judging from all reports of the time, the sex life of European men and women in the 1920s was far less restricted than it had been before the war. Then, for the first time in over a century, men and women could begin to experiment with, and play at, sex. Much of this innovation was accomplished only with the accompanying anxiety that besets pioneers and revolutionaries, but it did take place. The extent of the change is revealed by the exploration of the humorous aspects of sex in the work of D. H. Lawrence and Henry Miller.

Chapter 16

Crazy Cock
and Lady Jane

    A basic psychoanalytic rule is that passionate denials usually conceal strong antithetical sentiments—equally passionate, though repressed, affirmations. Clinical evidence shows that this theorem applies to the fanatical protector of public morality who denounces feelings which he can only allow himself to enjoy in the disguise of smut hunting and paroxysms of moral indignation. For the historian, the activity of the censor provides a good index to the morality that society *thinks* it can tolerate. Accordingly, the more active the vice squads and the censors, the greater the turbulence in the social unconscious. Since modern societies for the suppression of vice flowered during the Victorian period, we may conclude that underneath the composed exterior seethed vast stores of repressed sexuality. Freud identified this subterranean ferment, which first cracked a smile in the works of D. H. Lawrence and then bubbled over with Henry Miller's hearty laughter.

    The action of government prosecutors and private societies for the suppression of vice confirms my general argument that since the 1880s there was a steady relaxation of sexual repression in Europe and America. Today it is difficult to believe that Flaubert was taken to court on charges of corrupting public morals for the publication of *Madame Bovary* in 1857. The government prosecutor objected to two scenes of seduction which contained no explicit sexuality, Emma Bovary's passing interest in religion between her two adulterous affairs, and the vivid description of her suicide from arsenic poisoning. Flaubert was acquitted by the court, but the fact that this mild story of the hapless affairs of the wife of a provincial doctor could attract the attention of a public prosecutor in 1857 indicates how limited the range of socially acceptable behavior was at that time.

Other acts of suppression chronicle the changes in sexual morality that have taken place since Flaubert. The censorship of Zola's *Earth* in 1888 and Joyce's *Ulysses* in 1930 illustrate the varying standards of socially acceptable sexual activity. Americans struggled under the watchful eye of Anthony Comstock, the dry-goods clerk who made his way up through the ranks in the YMCA to become the chief agent for the New York Society for the Suppression of Vice in 1872. In 1913 he quantified his accomplishments with an extraordinary boast: "In the forty-one years I have been here I have convicted persons enough to fill a passenger train of sixty-one coaches, sixty coaches containing sixty passengers each, and the sixty-first almost full. I have destroyed 160 tons of obscene literature."[1] The story of one unfortunate victim of this era of sexual repression in America was reported in a German periodical for sexual reform in 1907. Somehow the editors uncovered the too-good-to-be-true story of a young man who was rejected from an army recruiting center in Joplin, Missouri, because he had a naked Eve tattooed on his chest. Two days later he returned, having had a dress tattooed on her, and was accepted.[2] The history of censorship until World War I is a strange mixture of bizarre humor and profound tragedy, for the Comstock laws prohibited distribution of manuals on sex education and contraception along with hard-core pornography and therefore intensified the already destructive effect of Victorian prudery by the added threat of criminal prosecution.

The war knocked much of the fragility out of the European and American moral conscience, so that when the authorities again attempted to quash the distribution of obscene literature in the 1920s they found more daring subjects for attack. Two works that brought out the guardians of public morality in the postwar years show how much that morality had changed since the days of Anthony Comstock. Before the war *Lady Chatterley's Lover* and *Tropic* of *Cancer* would have been condemned and suppressed as obscene with little opposition. As it was, both books labored for years under the stigma of censorship and emerged for widespread public attention only after a long struggle. In addition to using the forbidden four-letter words and describing sexual and other traditionally taboo bodily processes in vivid detail, they explored the humorous nature of sex. That quality unites these works and constitutes a major addition to the popular view of sexuality.

Lady Chatterley's Lover is one of several works in which D. H. Lawrence explored the sexual problems of his age. *Sons and Lovers* (1913) was about the Oedipal conflicts of a young man sexually paralyzed by, and bound to, his mother. *Women in Love* (1920) concerned tensions and rivalries among two adult couples struggling with the changing sexual mores of the war years, and *The Fox* (1923) investigated a tortuous lesbian affair. *Lady Chatterley's Lover* is a sexual *Bildungsroman* in which Lawrence reconstructs the history of female sexual problems from Connie's early education, which led her to fear that sex would enslave her to a man, through her sexually frustrating marriage, to her ultimate fulfillment in a sexual union with Mellors.

Constance Chatterley has been brought up to believe that sex is sordid and leads to a woman's subjugation: men "insist on the sex thing like dogs." The one way out is to learn to yield to a man without really giving herself. "Rather she could use this sex thing to have power over him. For she only had to hold herself back in sexual intercourse, and let him finish and expend himself without herself coming to the crisis: and then she could prolong the connection and achieve her orgasm and her crisis while he was merely her tool." [3] She marries Clifford Chatterley, a cerebral man for whom sex is "merely an accident." His disinclination for sex becomes a permanent disability as a result of his paralysis from a wound received in World War I. Soon after his return Lady Chatterley begins to crave sex and has an affair with Michaelis, but his "pathetic, two-second spasms" leave her unsatisfied. Their affair is a sexual power struggle. Following one mutually unsatisfactory attempt he challenges her: "You couldn't go off at the same time as a man, could you? You'd have to bring yourself off!" Lawrence intrudes to comment on the absurdity of Michaelis's complaint, because "like so many modern men, he was finished almost before he had begun. And that forced the woman to be active."[4] Their quarreling over the proper active and passive roles and the problem of timing orgasms for mutual fulfillment echoes the formal discussion of these problems in the sex manuals of the period. Emil Paulk, as I have noted, speculated that during the war European men were perhaps for the first time learning to wait and help women have their climax. Michaelis does not wait for Connie, but their argument on the subject suggests that Lawrence believed that it would be appropriate for a couple at least to discuss the problem at that time.

There is some validity to the argument of a recent critic that Lady Chatterley's fulfillment in the course of her affair with Mellors subtly tends to reinforce the traditional view that women are slaves to men, despite Lawrence's repeated assertions to the contrary.[5] The argument, and it is insightful, is that even though Lawrence portrays the sexual liberation of a repressed Victorian woman, the process of her liberation, explained as a consequence of her sexual relations with the gamekeeper, implies that women are still totally dependent on men for their happiness. The only change has been that in place of the traditional political and economic subjection of women there now exists a sexual subjection, because women are so vitally dependent on an erect penis for their pleasure. But to limit one's evaluation to this one point is to miss the broader historical importance of the novel and its comment on human existence, both male and female.

The sexual relationship between Connie and Mellors is a celebration of sexuality in general, not just male sexuality. From the title we are led to think that the book is about Lady Chatterley's lover, but it is really about the love life of Lady Chatterley. Her marriage to Clifford offers no sexual release, and to aggravate the frustration of his sexual impotence, Clifford takes satisfaction from debunking the pleasures of the body. One afternoon Connie comes upon Mellors stripped to the waist, bathing himself on the estate grounds. Her pleasure from this visual experience is a prelude to the orgy of sensuousness that follows.

The more explicit moralizing about sex occurs during several parties given by Clifford, when sexual conservatism and sexual liberation confront one another in the persons of Lady Bennerley and Tommy Dukes. In response to Lady Bennerley's claim that happiness requires forgetting the body, Tommy Dukes argues that the only bridge across the chasm that confronts civilization is the phallus. "Give me the resurrection of the body!" he says, as Connie reflects to herself, "Give me the democracy of touch, the resurrection of the body!" The democracy of touch would enable her to transcend the class lines that separate her from Mellors, a gamekeeper, and enjoy the pleasures of the body with him. As the story develops, her response to him becomes progressively more spontaneous and more complete. Consider one description of her sexual experience.

> And it seemed she was like the sea, nothing but dark waves rising and heaving, heaving with a great swell, so that slowly her whole darkness was in motion, and she was ocean rolling its dark, dumb mass. Oh, and far down inside her the deeps parted and rolled asunder, in long, far-traveling billows, and ever, at the quick of her, the depths parted and rolled asunder, from the centre of soft plunging, as the plunger went deeper and deeper, touching lower, as she was deeper and deeper and deeper disclosed, and heavier the billows of her rolled away to some shore, uncovering her, and closer and closer plunged the palpable unknown, and further and further rolled the waves of herself away from herself, heaving her, till suddenly, in a soft, shuddering convulsion, the quick of all her plasm was touched, she knew herself touched, the consummation was upon her, and she was gone. She was gone, she was not, and she was born: a woman.
>
> Ah, too lovely, too lovely! in the ebbing she realised all the loveliness. Now all her body clung with tender love to the unknown man, and blinding to the wilting penis, as it so tenderly, frailly, unknowingly withdrew, after the fierce thrust of its potency. As it drew out and left her body, the secret, sensitive thing, she gave an unconscious cry of pure loss, and she tried to put it back. It had been so perfect! And she loved it so! [6]

This kind of explicit discussion of a woman's experience of sexual intercourse had generally been restricted to pornography, but Lawrence's literary credentials were well established by 1928, and the critics were forced to deal with this book that also examined a number of other topics with equal candor.

Though much of the discussion of sexuality is dominated by a deadly seriousness—at times Lawrence seems to imply that the fate of modern civilization rests on the resurrection of the body—there are some interludes when he touches on the comical aspect of sex. The ability to explore the humorous side of sexuality constitutes a significant stage in the progress of the European sexual consciousness. In 1905 Freud had argued that laughter at certain jokes comes from the release of energy when repressed sexual instincts and fantasies emerge into consciousness. But Freud never acknowledged that human sexuality *per se* was humorous or ridiculous. Twice Lady Chatterley observes the "supremely ridiculous" aspect of Mellors's movements during the sex act. "The butting of his haunches seemed ridiculous to her, and the sort of anxiety of his penis to come to its little evacuating crisis seemed farcical. Yes, this was love, this ridiculous bouncing of the buttocks, and the wilting of the poor insignificant, moist little penis." [7] Mellors introduces the euphemisms "John Thomas" and "Lady Jane" to enable him to discuss his sexuality with Connie free from either the clinical or the vulgar connotations of conventional terms. John Thomas takes on an identity of his own and on occasion appears to exist apart from Mellors. With these terms and the heavy cockney accent he uses to further lighten the mood, Mellors is able to communicate his sexual feelings and needs to Connie in a good-natured manner. "Tha ma'es not o' me, John Thomas. Art Boss? of me? Eh well, tha'rt more cocky than me, an' tha says less. John Thomas! Dost want *her?* Does want my Lady Jane? Tha's dipped me in again, tha hast. Ay, an' tha comes up smilin'.—Ax 'er then! Ax Lady Jane! Say: Lift up your heads o' ye gates, that the king of glory may come in. . . . Cunt, that's what tha'rt after. Tell Lady Jane tha wants cunt. John Thomas, an' th' cunt o' Lady Jane!" [8]

The sexual liberation of Connie and Mellors allows them playfully to explore each other's bodies. Connie weaves flowers into the hair on Mellors's chest, and then into his pubic hair. Another

11. Gustave Courbet, *La Source*, Paris, Musée du Louvre, courtesy Caisse Nationale des Monuments Historiques.

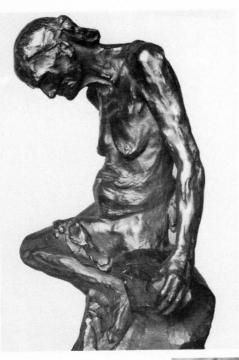

12. Francois Rodin, *La Belle Heaulmière*, Musée Rodin, Paris, courtesy photo Musée Rodin.

13. Henri de Toulouse-Lautrec, *Femme tirant son bas*, Paris, Musée du Louvre, courtesy Cliché des Musées Nationaux.

14. Georges Rouault, *Nu aux jarretières roses*, Paris, Musée d'Art Moderne, courtesy Photographie Bulloz.

15. Pablo Picasso, *Les Demoiselles d'Avignon*. Oil on canvas, 8' by 7' 8". Collection, The Museum of Modern Art, New York. Acquired through the Lillie P. Bliss Bequest.

16. Paul Cézanne, *The Bathers*, London, National Gallery.

17. Egon Schiele, *Sitzender Akt mit aufgestützten Armen*, Vienna, courtesy Lichtbildwerkstätte 'Alpenland.'

18. Egon Schiele, *Umarmung*, Vienna, Osterreichische Galerie.

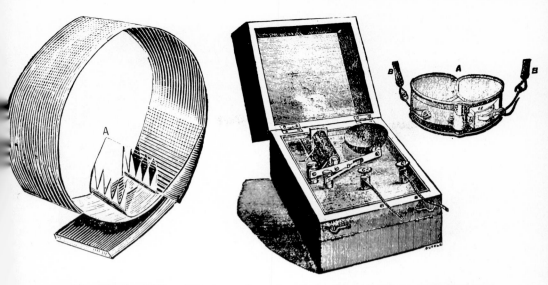

19. The Urethral Ring (Photograph 19a) and the Electric Alarum (Photograph 19b), used by some men to prevent nocturnal emissions. Reproduced from ON THE PATHOLOGY AND TREATMENT OF GONORRHOEA AND SPERMATORRHOEA by John Laws Milton, New York, 1887.

20. The "Geradehalter" used to force children to maintain an erect posture. Reproduced from DAS BUCH DER ERZIEHUNG AN LEIB UND SEELE by D. G. M. Schreber, 3rd edition, Leipzig, 1882.

21. The "Kopfhalter" forced the child's head to remain in an upright position. Reproduced from DAS BUCH DER ERZIEHUNG AN LEIB UND SEELE by D. G. M. Schreber, 3rd edition, Leipzig, 1882.

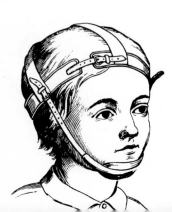

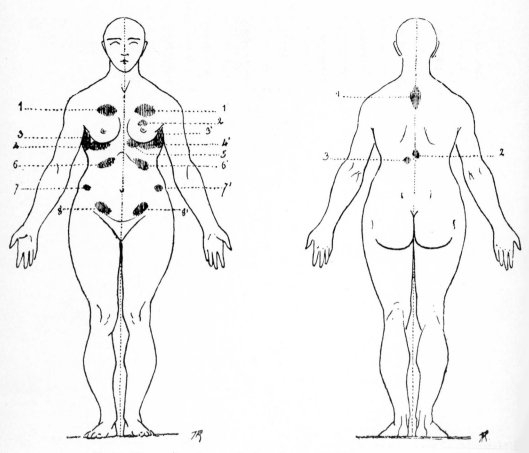

22. Jean Charcot's chart of the "hysterogenic zones," which when stimulated generated conversion forms of hysteria in certain patients. Reproduced from LEÇONS SUR LES MALADIES DU SYSTÈME NERVEUX by Jean Charcot, Paris, 1872.

time her hand wanders over his back and settles on his genitals. "And the strange weight of his balls between his legs! . . . What a strange heavy weight of mystery, that could lie soft and heavy in one's hand!" [9] After another sexual embrace his fingertips touch "the two secret openings to her body," and he says: "An' if tha shits an' if tha pisses, I'm glad. I don't want a woman as couldna shit nor piss. . . . Tha'rt real, tha art! Tha'rt real, even a bit of a bitch. Here tha shits an' here tha pisses: an' I lay my hand on 'em both an' like thee for it." [10] In an afterword to the novel Lawrence comments on the significance of this scene. "In the past, man was too weak-minded, or crude-minded, to contemplate his own physical body and physical functions, without getting all messed up with physical reactions that overpowered him." [11] Lawrence speculates that the insanity of Jonathan Swift was partly traceable to fear of bodily functions, in particular the excretory function. "In the poem to his mistress Celia, which has the maddened refrain, 'But—Celia, Celia, Celia s***s' (the word rhymes with spits), we see what can happen to a great mind when it falls into panic. A great wit like Swift could not see how ridiculous he made himself. Of course Celia s***s! Who doesn't? And how much the worse if she didn't." [12] Lawrence further explains that the purpose of the book is to enable men and women "to be able to think sex, fully, completely, honestly, and cleanly." [13] After her affair with Mellors, Connie finds herself responding differently to her father's body. For the first time in her life she notices her father's legs. "Connie woke up to the existence of legs. They became more important to her than faces, which are no longer very real. How few people had live, alert legs!" [14] Her thoughts are a subtle commentary on the long history of concealment of the legs of both men and women which prevailed throughout the nineteenth century and well into the twentieth.

The discovery of the body was part of a broad movement which Lawrence believed was at that time just getting under way, and in the course of the narrative he offered some theories about the historical significance of the developments of his time. "The human body is only just coming to life. With the Greeks it gave a lovely flicker, then Plato and Aristotle killed it, and Jesus finished it off. But now the body is coming really to life, it is really rising from the tomb." [15] In a subsequent short story Lawrence reworked the story of the ascension to improve upon Jesus's sexually repressive historical role.

Lawrence had originally intended to title his story of Jesus *The Escaped Cock*, but his publisher eventually chose the more conservative title *The Man Who Died*. It is interesting to note that Henry Miller had second thoughts about the title for his first novel, *Crazy Cock*, and called it *Tropic of Cancer*. The fact that both books were finally published under titles that concealed the central focus on sexual liberation suggests that both can be viewed (to borrow a phrase from Freud) as "compromise formations" between their impulse to challenge the restrictive sexual morality of the age and the force of that morality as imposed by artistic standards and public censorship. I have taken that common feature of their work as a basis for considering their work together in this chapter. Each, in his own way, coaxed modern man to allow his sexuality to run free for a time, to go "crazy" and "escape" the restrictions that had suppressed it for ages.

Lawrence's original title refers to the fate of a proud cock that Jesus liberates from his bondage and finally places in a new roost where he can fulfill his proud function without the hated tether that had strangled his masculinity in his first home. Jesus then wanders off to find the life forces that are dead in him. He is discovered by the Egyptian goddess Isis, who has searched for her dead lover Osiris, whose body was torn apart and scattered over the world. She has assembled everything save his penis—"the last reality, the final clue to him"—when Jesus appears as a lost traveler. She believes that he will supply the missing part which she must bring to life to fecundate her womb. Jesus is hesitant because the human touch recalls to him the painful wounds inflicted upon him at his crucifixion. The incantations of the Egyptian goddess finally overcome the asceticism of Christianity, and the man who died comes to life. This time his resurrection is in the flesh. "He crouched to her, and he felt the blaze of his manhood and his power rise up in his loins, magnificent. 'I am risen!' " [16] The long-awaited resurrection, Lawrence implies, involves the resurrection of the body and is symbolized in the allegory by Jesus's erection. After two millennia of asceticism the Christian world is ready for a fully potent messiah who is no longer afraid of human contact.

Though there are traces of humor in Lawrence's treatment of sexuality, much of his work is deadly serious and strongly moralistic. He allowed himself a few jocular remarks about the human body and the ridiculousness of the aesthetics of the sex act, but it was only in the work of Henry Miller that Western literature got a generous measure of the Rabelaisian tradition that had been so muted throughout the Victorian period.

The opening paragraphs of *Tropic of Cancer* (1934) announce Miller's love of art, his passion for food, and his lust for his friend's wife, Tania. While the world is being destroyed by a cancerous life-denying spirit, he proclaims his extravagant life style which, he is determined, shall consume all of his energies and give uncensored expression to all of his fantasies. The first of these fantasies reveals the fury and the futility of his sexual desire. His lusty prose recreates the mental images that Freud dug out of his patients' dreams and fantasies in the formal setting of psychoanalysis. Miller gives poetic expression to this primary process of the human mind.

> O Tania, where now is that warm cunt of yours, those fat, heavy garters, those soft, bulging thighs? There is a bone in my prick six inches long. I will ream out every wrinkle in your cunt, Tania, big with seed. I will send you home to your Sylvester with an ache in your belly and your womb turned inside out. Your Sylvester! Yes, he knows how to build a flame, but I know how to inflame a cunt. I shoot hot bolts into you, Tania, I make your ovaries incandescent. Your Sylvester is a little jealous now? He feels something, does he? He feels the remnants of my big prick. I have set the shores a little wider, I have ironed out the wrinkles. After me you can take on stallions, bulls, rams, drakes, St. Bernards. You can stuff toads, bats, lizards up your rectum. You can shit arpeggios if you like, or string a zither across your navel. I am fucking you, Tania, so that you'll stay fucked.[17]

To view this fantasy as degrading to women is to fail to see that it is essentially a transparent confession of the basic fears that trouble modern man—sexual rivalry, fear that his penis is not large enough, and the challenge of woman's superior sexual capacity. He dares to give a precise measurement of his erect penis, thereby allowing comparison with other men who always loom larger in the imagination. His fantasies of permanently reaming out her cunt, filling it up, and leaving it aglow so that Sylvester will know that someone has been there before all derive from fears that somehow he will not be big enough and that the organic memory of his presence will fade all too quickly. He returns to this central concern with the final, absurd claim that he will fuck Tania so that she will stay fucked. No woman can "stay fucked." However vigorously the male stakes his claim, the passage of time invariably returns the female genitals to a quiescent state, once again wanting a penis.

The prostitutes that inhabit Miller's Paris underworld of the early 1930s are his main source of sexual contact, and they embody both the evils of bourgeois society and the primitive human sensuousness that gets lost in it. His experiences with them inspire long monologues on human sexuality. One of his prostitutes, Germaine, has a special reverence for the sex that she sells to anyone for a few pieces of silver. He loved the way she would straddle a *bidet* and soap herself while nonchalantly chatting with him and then approach "rubbing her pussy affectionately, caressing it, patting it." [18] With Germaine he revives the figure of the happy prostitute who enjoys her work because of her need for sex. She needed a man—any man—"with something between his legs that could . . . make her grab that bushy twat of hers with both hands and rub it joyfully, boastfully, proudly, with a sense of connection, a sense of life." Germaine is the totally sexual female animal who comes to life only during the sex act. "That was the only place where she experienced any life—down there where she clutched herself with both hands." [19] The portrait of Germaine is presented, of course, from a man's point of view. It is not intended as a full account of the female psyche or even of female sexuality, but as a confession of the more absurd and self-deceiving contents of the male ego. No woman exists solely between her legs, but at times men like to believe they do to give the sex act a completeness that it rarely has, especially with prostitutes. Henry Miller was not boasting that Germaine went into ecstasy when they had sex, because he knew that the careful reader would know that she did not. He exaggerated his account just enough to draw out the reader's fantasies and confront him with them. He is mocking and laughing and is convinced that men are foolish to think that women fulminate and lose control of their senses when a man touches them.

With *Tropic of Cancer* the European reading public finally got the full confession promised by Rousseau but never delivered. Though changing literary and ethical conventions partly account for the many differences, Rousseau's *Confessions* appear tight-lipped compared with the openness of Henry Miller's autobiographical fantasies. There is a secretive quality to Rousseau even when he is making his precious few confessions. Miller openly displayed the workings of his sexual fantasies, and, I would speculate, the censors banned it to protect the public conscience from being exposed to such an explicit reminder of its own repressed contents.

Miller explored a full range of sexual experience from tragedy to farce. While Lawrence's account of the humorous side of sexuality was limited to Connie's observations about Mellors's thrusting, Miller revels in imaginative orgies of sexual burlesque. The story of a trip to a brothel with a Hindu friend becomes a good-natured comment on the disastrous consequences of cultural differences in the world of sex. Miller and his friend each pick a prostitute and go off into separate rooms. Before going into the room, his friend asks directions to the toilet, and, Miller wrote, "not thinking it was anything serious I urged him to do it in the *bidet*." Soon he heard the girl in the next room calling his friend a dirty pig. Then the madame arrived and began yelling. When Miller entered his friend's room he was with a group of prostitutes staring into the *bidet* where there were "two enormous turds floating in the water." The madame put a towel over it and shouted "Frightful! Frightful! . . . Never have I seen anything like this!" That debacle becomes the starting point for a reflection about the human condition. "Everything is endured—disgrace, humiliation, poverty, war, crime, *ennui*—in the belief that overnight something will occur, a miracle. . . . And so I think what a miracle it would be if this miracle which man attends eternally should turn out to be nothing more than these two enormous turds which the faithful disciple dropped into the *bidet*." [20] The story is an invitation to enjoy the comical aspects of sexuality. Miller's probing wit lays open the shadow world of the masculine ego—its pride, its vanity, and the ludicrous insults and failures that continually compromise it.

Through the ramblings of his American friend, Van Norden, Miller identifies the ultimate elusive goal of male sexual curiosity. For over thirty pages Van Norden tells Miller about his various "cunts"—his "Danish cunt," "old cunt," "a cunt by the name of Norma," and "the rich cunt Irene." Van Norden's "Georgia cunt" had been pestering him recently, and the last time they went to bed he discovered that she had shaved her pubic hair. She had "shaved it clean . . . not a speck of hair on it." After a lifetime of hunting he suddenly discovered that he was repelled by a woman's vagina. It appeared to him "like a dead clam or something." But curiosity soon displaced his disgust and he set about investigating, lighting it with a flashlight. He interrupted his story by reflecting on the significance of his discovery. "You get all burned up about nothing . . . about a crack with hair on it, or without hair. It's so absolutely meaningless that it fascinated me to look at it. I must have studied it for ten minutes or more. When you look at it that way, sort of detached like, you get funny notions in your head. All that mystery about sex and then you discover that it's nothing—just a blank. Wouldn't it be funny if you found a harmonica inside . . . or a calendar? But there's nothing there . . . nothing at all. It's disgusting. It almost drove me mad." [21] Van Norden's research and commentary explode the great secret mystery of the female sex organs. The American in Paris discovers the absurdity of his life's sexual cravings by examining the shaved vagina of his "Georgia cunt."

To the vast collections of philosophical treatises on sex, the clinical reports, and the poetic eulogies, Miller adds his self-revealing autobiographical fantasies, which focus alternately on the farcical and on the sublime. Amidst the picaresque clowning there emerges a serious moral tone reminiscent of Lawrence. The dialectic of mind and body oscillate in Miller between sensuous experience and philosophical rumination. Unreflective carnality eventually inspires artistic activity, which for him takes the form of an imaginative mixture of fantasy and memory. And when long contemplation threatens to remove him from the corporeal, he returns to the living embraces of the endless stream of women who provide the living touch he needs to restore his sensuous involvement in life. "At the extreme limits of his spiritual being man finds himself again naked as a savage. When he finds God, as it were, he has been picked clean: he is a skeleton. One must burrow into life again in order to put on flesh. The word must become flesh." [22] And so he returns to his whores again and again in order for the word to become flesh, and then those strained embraces begin once again to seed the workings of his poetic imagination.

Late one night he and his friend Fillmore pick up two whores and bring them back to his room. As they all undress and he becomes lost in thought, one of the whores wraps her thighs around his head and he finds himself staring at her vagina. "Suddenly I see a dark, hairy crack in front of me set in a bright, polished billiard ball: the legs are holding me like a pair of scissors. A glance at that dark, unstitched wound and a deep fissure in my brain opens up." [23] His thoughts begin to take shape as a series of images about human existence. "I see again the great sprawling mother of Picasso. . . . Molly Bloom lying on a dirty mattress for eternity. . . . Wild, wild, utterly uncontrollable laughter, and that crack laughing at me too, laughing through the mossy whiskers." He compares his thoughts with Van Norden's. "When I look down into that crack I see an equation sign, the world at balance, a world reduced to zero. . . . Not the zero on which Van Norden turned his flashlight, not the empty crack of the prematurely disillusioned man, but an Arabian zero rather, the sign from which spring endless mathematical worlds, the fulcrum which balances the stars. . . . When I look down into this fucked-out cunt of a whore I feel the whole world beneath me, a world tottering and crumbling." He then considers the static quality of life about him. "In the four hundred years since the last devouring soul appeared, the last man to know the meaning of ecstasy, there has been a constant and steady decline of man in art, in thought, in action. The world is pooped out: there isn't a dry fart left. . . . The dry-fucked-out crater is obscene. More obscene than anything is inertia. More blasphemous than the bloodiest oath is paralysis." This denunciation of the dry and inert quality of life leads to a statement of his positive philosophy: "Do anything, but let it yield ecstasy." He loves everything that flows: "rivers, sewers, lava, semen, blood, bile, words, sentences . . . the amniotic fluid when it spills out of the bag . . . even the menstrual flow that carries away the seed unfecund." [24] That whore's "fucked-out crater" symbolizes for him both life and death. With his head locked between her thighs, he is forced to confront that great zero which is the beginning and the end of all things. In that moment

Miller understands and accepts the elusive goal of male sexual desire, the mysterious place from which life comes. But it is soiled by experience—by the cash nexus of human relations, by sexual repression, and by self-deception—all necessary to screen the frail masculine ego from the truth about itself. It inspires uncontrollable laughter and the gloomy vision of an arid and inert existence.

Henry Miller was the first serious writer of the modern age to write an honest account of his sex life—its pathetic failures, its occasional triumphs, and its endless disillusions. His driving passion to discover all led to this revealing confrontation with himself, face to face with a whore's "fucked-out cunt." In that moment male sexual curiosity and the striving that it inspired came to a brief rest, the veil of mystery was stripped away, and his thought could once again begin to take flight and consider the loftier possibilities of human sexuality. With an unerring feeling for the essential, Miller finally accepted the comical and yet deadly serious nature of the object of male sexual desire, that beguiling combination of the beautiful and the ugly that had so mystified generations of men who preceded him.

Chapter 17

Body Politics in Germany

The sexual revolution was not an uninterrupted process of liberation and relaxation. While Lawrence and Miller were moralizing on behalf of sexual freedom, an intense racism began to penetrate the physical culture movement in Germany.

Though Germany emerged from the wars of unification the most formidable military power in Europe, the German people lacked the strong sense of cultural unity that the English and French had enjoyed for centuries.[1] Waves of immigrants from Eastern Europe and Russia in the 1880s and then again around 1917 added to the already disparate ethnic groups that made up the Reich and generated a strong chauvinism among those who considered themselves the rightful inhabitants of the fatherland. German *völkisch* philosophy dwelled on the physical and spiritual beauty of the German race and denounced its incipient contamination by those strange refugees from Eastern Europe with their un-German languages, foods, clothes, and, in the case of the Jews, Sabbath. The differences did not, in clashing, forge a higher synthesis, but tended rather to foster resentment among those who longed for national unity. Throughout the nineteenth century German thought shifted dramatically between positive and negative views about the cultivation of the body, its free representation in art, and its proper regulation in the service of national interests.

The Nazi philosophy of physical culture was an unstable amalgam of bombastic laudations of the well-developed body and panicky condemnations of sexual freedom.

From its inception the physical culture movement was linked with military preparedness, national survival, and the extirpation of foreign elements from German life. In the wake of defeat by Napoleon's Grand Army in 1806 Germans searched for ways to develop themselves into better soldiers. Friedrich Jahn revived the Hellenic ideal of "a sound mind in a sound body" and appealed to Germans to develop their bodies by systematic exercise. His gymnastic societies became centers for German nationalism where young men could learn to coordinate their movements into a unified and disciplined display of physical excellence.[2] German gymnastics contrasted with the kind of physical education that later developed in France and England, where more recreational team sports became popular. As one historian of sports has observed: "The *völkisch* gymnasts . . . were deadly in earnest. They were organized into quasi-military hierarchies and formed paramilitary clubs that were meant to prepare the German *Volk* for a test against its enemies who were preventing German political self-realization."[3] Notwithstanding its military emphasis and its persistent underlying concern to protect Germans from "contamination" by contact with other races, the physical culture movement provided a solid theoretical and institutional foundation for the cultivation of physical excellence during a century notorious for the neglect of the human body. The movements in support of clothing reform, sun bathing, and physical education of the late nineteenth century shared this aim. Their central concern was to restore the natural man who was being destroyed by high fashion, city life, and sedentary work.

One embarrassing challenge to the belief in the physical superiority of the Germans occurred during the Olympic Games held in Berlin in 1936. A century of racist thinking was refuted in 10.2 seconds when the black American sprinter Jesse Owens won the 100-meter dash and set a new world record. "The fastest man alive" was a Negro, and German theoreticians had to work hard to reconcile their pronouncements on the inferiority of the darker races with the fact that Jesse Owens had won four gold medals. Films of the games show Hitler wincing each time Owens left a field of white athletes behind. When in 1938 Leni Riefenstahl finally released her cinematographic chronicle of the Berlin Games, *Olympia*, her fascination with Owens's black body and graceful movement was apparent, and the Führer was outraged.

The history of art criticism in Germany reveals a similar mixture of the inspirations and fears that guided the physical culture movement, although the anti-Semitic element appeared in German aesthetics more toward the end of the century. Since 1887 Ferdinand Avenarius had warned Germans about the dangers of the Jewish influence in art in his monthly publication *Art Guardian (Kunstwart)*. During the early decades of the twentieth century the anthropologist Adolf Bartels became the custodian for pure *völkisch* art in Germany. In 1925 Bartels complained that the democratic ideals of the Weimar Republic and their manifestation in Expressionist art represented a pernicious Jewish influence on German culture.[4] Hans F. K. Gunther continued to pursue the subject in *Race and Style* in 1926, which decried the artistic influence of the "lower races."

Nazi suppression of foreign art was institutionalized with the League of Struggle for German Culture, founded in 1929 by Alfred Rosenberg. Rosenberg enlisted the assistance of Bartels and Paul Schultze-Naumberg to publicize the League's message. In the following year Schultze-Naumberg was appointed director of the State Academy of Art in Weimar with the task of creating a center to promote German art, and in 1931 he began a lecture tour to expose the dangers of "degenerate art." He developed a method of propaganda which involved the juxtaposition of the paintings of German Expressionists, particularly the distorted nude figures, with photographs of cretins, cripples, and the congenitally deformed. The career of Schultze-Naumberg embodied both the progressive and chauvinistic elements of the physical culture movement. Around the turn of the century he campaigned for the acceptance of clothing for women that would follow the natural lines of the body and prevent the disfiguring from corsets. Over the years he was influenced by the racist anthropology of Gunther and Walther Darré. He shifted easily from campaigning on behalf of "natural clothing" to supporting art fashioned after natural (i.e., healthy and German) bodies. In *Art and Race* (1928) he complained that the preoccupation of contemporary artists with idiots, prostitutes, and "sagging breasts" (*Hängebrust*) was corrupting German aesthetic sensibilities and morals and creating a visual image of German society as a "hell of *Untermenschen*." In the early 1930s he gave an anti-Semitic twist to his aesthetic theory to bring it into line with the needs of the Nazi party. When Frick appointed him head of the League of Struggle for German Art, Schultze-Naumberg could boast a background of dedication to the cause of protecting the "natural" man from the nefarious influences of urban intellectuals, Parisian fashions, Jewish tailors, and Bolshevik art.

Alfred Rosenberg traced the contours of a uniquely "Northern beauty ideal" in his monumental cultural hodgepodge *The Myth of the Twentieth Century* (1930), which provided theoretical foundation for Schultze-Naumberg's lectures and exhibits. That unique beauty, Rosenberg argued, is most vividly revealed in the face; the formation of the eyes, nose, mouth, and chin expresses the will and character of a people. His Nordic aesthetic contrasts sharply with the aesthetic reflected in the "typical" Jewish face—"crooked nose, hanging lips, piercing dark eyes, and wooly hair." These Jewish features were contaminating German art and creating a "racial chaos" of Germans, Jews, and foreigners. The Dutchman Van Gogh went insane and the Frenchman Gauguin died a leper, and the art that had sprung from these two foreign influences was a disease of the mind and body. German Expressionism in particular was dominated by Jews and Marxists. A new aesthetic was needed to extol the Northern racial ideal of beauty.

The defenders of German culture were active in 1930. In that year Gunther was given a chair in the "science of human breeding" at the University of Jena amidst a storm of protest by students and faculty. Hitler attended his inaugural lecture entitled "Biological Considerations of Corporeal Beauty," which warned of corruption of German art by racial *Untermenschen*. The censors suppressed a production of Brecht's *Threepenny Opera* and ordered the removal of a number of Expressionist paintings from a Weimar museum.[5]

For four years following its formation in 1933 the Reich Chamber of Culture under Goebbels's direction suppressed art that did not conform to Nazi cultural needs. Then on July 18, 1937, Hitler and Goebbels decided to bring public opinion into line on a large scale with a much publicized "Exhibition of Degenerate Art" at the House of German Art in Munich. Schultze-Naumberg's comparative method was used to illustrate "degenerate" Expressionist art. Placards placed over the groupings of paintings and photographs to ensure the proper response read: "This is how sick minds regard Nature," or "German peasants seen through Yiddish eyes." [6] Hitler opened the exhibition with a lecture condemning art inspired by primitive, Bolshevik, and Jewish minds which produced only "misformed cripples and cretins, women who inspire disgust, and men who are more like wild beasts." [7] The following day he opened a contrasting exhibition of "true" German art. It was organized by Adolf Ziegler, a painter whose realistic nude studies earned him the title "Master of the Pubic Hair." Ziegler depicted racially ideal types and set a model for officially sanctioned nudes for the remainder of the Nazi period. Just as landscape painters ought to paint German landscapes, so nude studies ought to depict the German body. The requirements for racial perfection eventually became so demanding that no living model could measure up to them, and so one popular Nazi sculptor, Arno Brecker, worked from a corporeal ideal he conjured up out of the racist tracts of Nazi theoreticians. Ironically, the traditional German preference for the ideal over the real recurred in Brecker's statues of superbly unreal "pure" German bodies.

The aesthetic and racial guidelines for Brecker's statues and other nude studies of the time were summarized in two writings of 1937. One by Schultze-Naumberg explained how to render the harmonious balance of grace and strength of the ideal German body. With a vividness reminiscent of the medical aesthetics of C. H. Stratz, Schultze-Naumberg in *Stylish Beauty* described the ideal Aryan breast. "The nipple had to be inverted and its areola flat, and the breasts themselves, pale pink with weak pigmentation, must sit full and well-rounded on the thorax." [8] Wolfgang Willrich's *Cleaning the Temple of Art* denounced the degenerate art held up for public ridicule at the exhibition of 1937. It began with the same "it's-about-time-we-did-something" tone that Hitler assumed in his opening speech for the exhibition. A degenerate rabble has taken over modern art, and the temple of art must be cleansed of disfigured nudes such as those by Nolde, Heckel, Schmidt-Rottluff, Kirchner, and Beckmann. To illustrate his argument he reproduced their paintings juxtaposed with clinical photographs of deformed bodies as Schultze-Naumberg had done. Willrich argued that Jews, Bolsheviks, and foreigners had produced an unnatural art that mocked human love with distorted faces and twisted bodies. Such work was created by sexual degenerates and must be suppressed, because it corrupted public morality. It employed models taken from the lower classes, and therefore could not portray an "ideal" Nordic beauty. The proper goal of art ought to be the development of "a deep, racial feeling" leading to a "purification of German blood." [9]

Willrich's art is a pictorial catalog of the "German" qualities forced into vogue by the Nazi overseers of culture. A girl blushes modestly as she fingers her long braids, a mother smiles winsomely upon the infant nursing at her milk-swollen breast. The men are muscular youths or wizened peasants with the ancient beauty of the German landscape carved on their craggy faces. All of Willrich's subjects are frozen in a stolid monumental calm—the antithesis of Expressionist animation. They are all well-behaved Germans, physically as alike as if they were members of one family, part of that pure, incestuous, racial unity that the Third Reich so fervently sought.

There is an ambivalent sensuousness in Willrich's work that is characteristic of a good deal of Nazi culture. He criticized Christian asceticism, which viewed the woman's body as a gate to hell. Germans, he wrote, needed to see a "healthy" portrayal of the body, not the Expressionists' flaunting of sexual pathology. But Willrich could not perceive that his own insistence on portraying only beautiful Aryans betokened a paralysis of the creative impulse and a fear of the great variety of bodies produced in a world of sexual freedom. The Nazi desire to create a master race incorporated the same contrasting elements of sensuousness and prudery that so limited the paintings of Willrich and inspired the stony, characterless muscle men and their matching mates sculpted by Brecker.

It must be remembered that in the nineteenth century, racial stereotypes provided some basic conceptual tools for comparative cultural studies and were part of the mainstream of anthropological and historical thinking. Darwin viewed racial variations as a product of natural selection along with the more marked differences between species, and Herbert Spencer uncritically applied European cultural standards to debunk the lower races. Though Hegel rejected the simplistic characterological studies of the phrenologists, he subscribed to a strong ethnocentrism with his untroubled dismissal of the Africans as historically insignificant. Even Marx occasionally indulged in some casual ethnic slurs when commenting on the backwardness of Asian and Indian civilizations. A great many nineteenth-century scholars studied physical variations of race to try to discover corresponding differences in the intellectual and moral sphere. They added to the vast literature on phrenology a variety of speculations about national and racial characteristics expressed in the body. Such racist thinkers as Spencer and Gobineau limited themselves to arguing for the superiority of European whites, while others became more aggressive and demanded enhancement of their presumed racial superiority. In England Karl Pearson extended Darwin and Spencer by insisting that it was the duty of the superior races to accelerate the pace of natural selection by inaugurating a eugenics program to "see that man is better and better born." [10] In *National Life and Character* (1893) Pearson warned Europeans about the menace of the black and yellow races and called upon whites to resist mongrelization. Gobineau had already argued in the 1850s that racial purity betokened greatness. To Gobineau's admiration for the pure Nordic races Vacher de Lapouge added anti-Semitism in the 1870s.[11] French racism in the 1890s was intensified by the fear of a sharply declining birth rate. In France as in England, however, racial thinking remained peripheral to major cultural developments, while in Germany the nagging search for a pure national heritage inspired widespread concern about the best way to ensure the preservation and higher development of the Aryan race.[12] In the 1850s German cultural spokesmen developed some elaborate theories about the unique properties of German minds and bodies. Wilhelm Riehl's *Land and People* (1857–1863) argued that the topography of a land influenced the body structures and facial expressions of its inhabitants. This paralleled the phrenological literature which sought to establish correlations between the shape of an individual's skull and his personality. Theodor Piderit's *Mimic and Physiognomy* (1858), Louis Gratiolet's *Physiognomy and Expressive Movements* (1865), and Darwin's *Expression of the Emotions in Man and Animals* (1872) all related body types to moral and intellectual capacities. Darwin's stature added respectability to this work, which abounded by the 1890s.

Characterology revived in Germany in the 1920s and contributed to the already massive literature on racial stereotypes. The studies of Ernst Kretchmer correlated personality with three basic body types—fat, thin, or muscular. In *Physique and Character* (1921) Kretchmer rejected earlier "naïve connections between bodily and psychic peculiarities," and he explicitly denied that any race had a preponderance of particular mental qualities. In 1932 a journal of phrenology, *Der Menschenkenner*, began a series of studies of eminent personalities. A lead article by the editor, Amandus Kupfer, gave a phrenological analysis of Hitler as a "deed and movement" type whose bones and muscles dominated his body. His fine hair signified "health, naturalness, and heartiness," and his uniquely styled moustache was designed to counter the strong American influence on facial aesthetics. In May 1933 the journal began to offer suggestions for breeding a superior race. Kretchmer's study of character types was far more sophisticated than the crude work of Kupfer, but to the larger population these studies appeared to support each other and gave the impression of a unified and intellectually respectable body of thought to the effect that there were racial types and that races were unequal. There was a common concern behind the phrenology of Gall and Spurzheim and the characterology of Kretchmer—the desire to find simple reasons for the complexities of individual human variation. And in Germany, where racial variations appeared to be so bewildering, these theories found a most receptive audience. Their popularity helps explain why Germans were so well disposed intellectually to accept the theory that Jews constituted a threat to the nation.

To understand anti-Semitism in Germany it is essential to look beyond the religious and social differences to the underlying fears and the sexual and biological theories they derived from. Around the turn of the century, when Jewish cultural life was exploding with creativity, boasting about German racial superiority became ever more frequently linked with the insistence on Jewish inferiority. The German aesthetician Christian H. Stratz studied standards of beauty among different races throughout the world. His analysis included some seemingly haphazard procedures such as measuring the distance between the nipples of the women of different races to determine which produced the most beautiful bodily proportions. He argued in *The Racial Beauty of Women* (1901) that the pinnacle of beauty was the blue-eyed blond found in Northern Europe. "Even the women of other races acknowledge the higher beauty of their lighter sisters and try to imitate them." [13] Pure German types were becoming tainted with undesirable racial attributes by inbreeding with those pushy representatives of the lower races—the Jews. He concluded with a warning about the imminent threat to that symbol of German superiority—the blond and blue-eyed woman, the most highly valued sexual object of the darker races. Stratz's warnings reveal a central anxiety and self-mistrust of some blond German men who believed that blond women were attracted to exotic dark-skinned men. We have seen that this fear surfaced during World War I, and in the postwar years it persisted in popular culture.

In 1905 the first issue of Lanz von Liebenfels's journal *Ostara*, subtitled *Library for Blonds and Supporters of Male Rights (Bücherei für Blonde und Mannesrechtler)*, warned Germans of the consequences of racial mixing. Over the years it helped develop the routine nineteenth-century anti-Semitism into the intense cultural paranoia that came to obsess Hitler and the Nazis. The journal catered to the underlying uncertainty that bordered on panic when certain Germans contemplated the dual threat of racial mixing and equal rights for women. *Ostara* was to awaken the blond race to its responsibility for preserving its purity. In 1907 Liebenfels identified the arch-enemy of the German race—the Jew. An article of 1908 speculated that the original sin was miscegenation. In 1910 he argued that a blond woman could be permanently contaminated from a single sexual contact with a man from another race. Some nineteenth-century biologists had claimed that a woman could be tainted with the blood of a man who impregnated her, but Liebenfels argued that semen could contaminate a woman even if she did not conceive a child. The higher races are in danger of becoming "apefied" *(veräfft)*. "The history of blond eroticism is inscribed on the body of the blond woman." She does not realize how she imperils the blond race by mixing with the lower races, and therefore it is the obligation of the blond knight to protect the race.[14] In 1913 Liebenfels prescribed some measures which could prevent the poisonous sexual touch of the lower "mishmash" people: sexual segregation, deportation, forced labor, sterilization, or even liquidation.[15]

The anti-Semitism we have observed in the aesthetics of Schultze-Naumberg and Willrich, the anthropology of Bartels and Gunther, and the racist tracts of Liebenfels shared in a central fear that the Jews imperiled the purity of the Aryan race. Toward the end of World War I this fear began to trouble a number of German thinkers, and through their work it began to concern ever wider circles among the German population. In 1928 Arthur Dinter's best-selling novel, *Sins Against the Blood*, warned of the dangers of miscegenation. Joseph Goebbels expressed his disgust with Jews in his first novel, *Michael* (1929). The hero complains that Jews made him "physically sick." "[The Jew] has raped our people, soiled our ideals, weakened the strength of the nation, corrupted morals. He is the poisonous eczema on the body of our nation."[16] Alfred Rosenberg (1930) viewed the Jews as "a kind of bacteria, a parasite that feeds on a larger body." He also recorded Weininger's characterization of Judaism as "an invisible cohesive web of slime fungus . . . spread over the entire earth."[17] Hermann Paull speculated about the effects of racial mixing on hereditary stock. Interracial love permits "the squandering of high-grade germ endowments," whereas monogamy "enables human reason to bring together high-grade hereditary stock for human breeding and to exterminate hereditary stocks of inferior grade." The Jews were clearly the richest source of "low-grade germ plasma."[18]

Modern genetics has concluded against Liebenfels's idea that semen can influence the hereditary make-up of a woman even if conception does not take place. But in the atmosphere of racial fanaticism of the 1930s some German thinkers became convinced that a single drop of venomous Jewish semen could forever destroy an Aryan woman's racial purity. The most influential proponent of this theory was the Gauleiter of Nuremberg, Julius Streicher, who developed it in a notorious article of 1935 in which he wrote: "Foreign albumen is the semen of the man of another race. During coitus it is absorbed by the uterus and passes into the blood. A single intercourse of a Jew with an Aryan woman suffices to poison her blood forever. She also absorbs the foreign soul along with the foreign semen."[19]

A corollary to this theory of Jewish sexual contamination

was the idea that syphilis was Jewish in origin. Lacking hard evidence to support this wild accusation, critics argued that the Jews' obsession with finding a cure implicated them in causing the disease.[20] The fact that Jewish scientists had monopolized the scientific study of sex and had made the major contributions to the fight against venereal disease seemed to establish their responsibility for its spread. Fritz Schaudinn, who first observed the syphilis spirochete, August Wassermann, who developed a test for it, and Paul Ehrlich, who discovered Salvarsan to treat it, were all Jewish. Some of the great pioneers of the study of sex, Otto Weininger, Eugen Steinach, and Friedrich Krauss, were Jewish, as were the first psychoanalysts, Freud, Adler, Stekel, Sachs, Reich, and Ferenczi.[21]

In *Mein Kampf* Hitler insisted that "all really significant symptoms of decay of the pre-War period can in the last analysis be reduced to racial causes."[22] His first reaction to the Jews he saw in Vienna in his youth, he maintained, was physical disgust. "Later," he recalled, "I often grew sick to my stomach from the smell of these caftan-wearers."[23] Hitler condemned the many ways Jews were corrupting the economy and politics of Germany, but he reserved his bitterest invective for racial mixing. He joined Liebenfels in viewing the original sin as miscegenation, and he too considered that the Jews were responsible for the spread of syphilis. "This Jewification of our spiritual life and mammonization of our mating instinct will sooner or later destroy our entire offspring."[24] Several historians have cited his fantasies about Jews seducing German women. "With satanic joy in his face, the black-haired Jewish youth lurks in wait for the unsuspecting girl whom he defiles with his blood, thus stealing her from her people."[25] In contemporary Germany, he complained, Jewish money could buy Aryan love. The natural union of healthy Germans was impeded by this "mammonization of love," and "foppish fashions" deftly concealed racial flaws and allowed "the seduction of hundreds of thousands of girls by bow-legged, repulsive Jewish bastards." The national interest required rather "that the most beautiful bodies should find one another, and so help to give the nation new beauty."[26] The proper function of a people's state must be "to see to it that only the healthy beget children [and to] declare unfit for propagation all who are in any way visibly sick or who have inherited a disease."[27]

When Hitler came to power in Germany he attempted to carry out this stated purpose and prevent the further "bastardization" of the German nation with a program to encourage parental pairing of Germans and to discourage, and eventually to forbid, sexual mixing of races.

Hitler began to regulate German sexuality within a month of his appointment as Chancellor. In February of 1933 the Nazis began a campaign to control venereal disease and outlawed prostitution (both problems linked by then with the Jewish menace), and in March a decree forbade nudism. In July the Nazis passed legislation to prevent congenitally diseased offspring by sterilizing persons suffering from incurable hereditary disabilities.[28] The Nuremberg Laws of September 1935 were conceived to protect German blood from foreign influences by prohibiting interracial marriages. They forbade marriages and extra-marital sexual relations between Jews and Germans. Jews were also prohibited from employing any female servant under forty-five years of age. A "Statute to Protect the Hereditary Health of the German People" of October 1935 allowed marriages only between those persons passing a medical and eugenics investigation, and it enabled authorities to sterilize habitual criminals and hereditarily defective individuals.[29] In October 1939 Hitler authorized physicians to "grant a merciful death" to the hopelessly sick.[30] In 1941 foreigners convicted of having had sexual relations with a German woman were subject to the death penalty, and the "final solution" of the Jewish question was put into operation in 1942. The fanatical concern for "cleaning" and "purifying" German thought, German art, and German blood culminated in this grisly slaughter to cleanse Europe of racial undesirables. A macabre footnote to the genocide is offered by the story of a professor of anatomy at the "Reich University" in Strassburg who wrote to Himmler in February 1942 requesting to be supplied with Jewish skeletons to fill out the university's collection from different races around the world. "The war in the east," he wrote, "offers us a chance to fill in this gap with the skeletons of Jewish-Bolshevik Commissars who are revolting, but characteristic, examples of the inferior races."[31] Himmler had his own more modest collection of skulls from different races.

At the same time as the Nazis were discouraging the

procreative activity of the racially undesirable, they were encouraging the reproduction of Aryans. In June 1933 the Nazis inaugurated a program of interest-free loans for married couples. A quarter of each loan could be forgiven for each child the couple produced. In 1936 Walther Darré organized the Fount of Life communities to care for unmarried women of "good blood" who bore children sired by members of the SS. In December 1938 Hitler inaugurated some awards for Aryan procreation. Four children earned a mother a bronze *Mutterkreuz*, while six, seven, or eight children were rewarded by silver or gold medals.

The Hitler Youth and the Strength Through Joy organizations in Germany gave particular emphasis to the cultivation of German bodies. The Reich Physician to the Hitler Youth addressed the young members in 1939 with a list of "Ten Commandments" which began with the statement: "Your body belongs to your nation." [32] This instruction made explicit the political regulation of sexuality that the Nazis had been assuming since their accession to power. The Strength Through Joy organizations were conceived to ease some of the pressure on German workers and soldiers in a joyful participation in working and fighting for the fatherland.

Nazi party rallies became ritual celebrations of German corporeal excellence. The rows of men and women performing mass calisthenics synchronized to lilting tunes displayed the many ingredients that had gone into the physical culture movement for over a century: the military inspiration of its founders, the robust enthusiasm of its followers, and the racial fanaticism of its Nazi overlords. The Nazis achieved an unprecedented degree of political control over the bodily existence of the German people. They demonstrated that the modern state could influence the way its citizens viewed their bodies and related to others' bodies. It inspired young men to steel their bodies for combat, and it persuaded young girls to produce babies for the Führer. It convinced large segments of the population that certain kinds of bodies were superior to others, and it marshalled enough support to establish death camps to destroy those bodies it believed racially threatening. The victims were tatooed, undressed, and then herded into the gas chambers like animals. Himmler knew that it would be easier to exterminate masses of naked bodies, whose human identity had been stripped away with their clothing.

One man living in Austria and Germany when the Nazis were coming to power was particularly well prepared to understand the anti-Semitism and the troubled sexual morality of the times. Wilhelm Reich was a Jew, a Communist, and a psychoanalyst, and his special interest in psychoanalysis was sexual theory. He embodied everything the Nazis most feared. His efforts to reconcile Marx and Freud and to explain the impact of an authoritarian state on sexuality earned him expulsion from the German Communist Party in 1933 and exclusion from the International Psychoanalytic Association in 1934. He went into exile in Denmark and Norway until he finally settled in the United States in 1938, where at first he found an understanding response to his ideas. In the 1950s he claimed that he had discovered the life-supporting energy of the universe—orgone—which he could capture and store in specially constructed Orgone Boxes. When he persisted in using Orgone Boxes to treat patients in violation of an injunction of the Food and Drug Administration, he was sent to Lewisberg Penitentiary, where he died in 1957, lost in a tangle of paranoid delusions deriving from a lifetime of struggle against the sexual morality that he found so destructive.

The story of Reich's martyrdom on behalf of sex reform began in 1919 when he became convinced that "sexuality [was] the center around which revolves the whole social life as well as the inner life of the individual." [33] In the following decade he studied Marx to understand the nature of that "social life," and Freud for the "inner life of the individual." In 1923 he made his first critical departure from Freud by introducing his own "orgasm theory." Until that time, he argued, "psychoanalysis knew only of an *ejaculative* and *erective* potency." According to Reich, men who are able to perform these functions experience only limited release of sexual tension during intercourse. Sexual health also involves "orgastic potency," which he defined as "the capacity for surrender to the flow of biological energy without any inhibition, the excitation through involuntary pleasurable contractions of the body." [34] For the next few years he searched for social origins of his patients' sexual problems, and then in 1927, after Freud refused to analyze him, he turned forthrightly to the study of Marxism. In 1928 he became active in the Austrian Social Democratic Party and attempted to bring his ideas on sex reform to a large number of people by opening some sex clinics for workers in Vienna, and then in Berlin, which disseminated information about birth control, abortion, and sexual technique. He concluded from interviews with his patients that the social alienation of the workers was manifested in their bodies. A rigid authoritarian society created rigid bodies. The workers' lack of privacy, hygiene, and education had depleted their capacity for sexual gratification. Their repressed character and rigid musculature blocked the free discharge of sexual tension during orgasm.[35]

This "character armor," Reich argued, was a corporeal expression of the "defenses" Freud observed in his patients. But while Freud conceived of these defenses as manifested primarily in the mind, Reich believed that the bodily manifestations were primary, and he developed a therapy to treat them directly. To replace Freud's "talking cure" Reich worked out a technique to liberate the pleasure-giving energies of the body. His therapy involved breathing and shouting exercises which were intended to free energy trapped in the body by muscular armor. He sought to relieve tension in the face, chest, or genitals by bringing about an uninhibited release of sexual energy accompanied by anger and tears. While progress in therapy could be observed by changes in breathing and facial expressions, the focus of the therapy was the capacity to experience a complete "orgasmic" discharge of sexual tension.

An article of 1932 traced the formation of the repressive morality that he believed prevailed in modern authoritarian societies.[36] His male patients from the Berlin working class revealed "unsure potency or erotic grossness," the women had "sexual incapacities or character deformations," and both sexes were unable to achieve full genital gratification. Reich attributed the cause of their misery to the greed, possessiveness and competition that prevailed in a capitalist society and which imposed a sex-negating morality through the patriarchal family. He concluded that the transition from matriarchy to patriarchy shifted the emphasis from sexual freedom to marital ties and from pre-marital sexual license to the demand for pre-marital chastity. It further generated a progressive division of society into classes. This class structure enabled the dominant class to impose whatever sexual morality best served its interests, and in a bourgeois-capitalist society there were two essentials—pre-marital chastity for the women of the upper classes and sexual repression for the working population. The repressive sexual morality of the bourgeois-capitalist society was imposed on the working class by the upper classes and then came to be self-imposed by the working class itself. Reich explained this process of internalization in Marxian style: Relations of production create classes with special interests and morals. The dominant class imposes its morality on all classes, and in time it dominates all social relations. Finally the lower classes fully accept this morality and begin actively to campaign on its behalf. "The exploited person affirms the economic order which guarantees his exploitation; the sexually repressed person affirms even the sexual order which restricts his gratification and makes him ill, and he wards off any system that might correspond to his needs." [37] This internalization, Reich believed, had occurred among the working classes in Germany and had made them responsive to the sex-negating morality of Nazism.

Reich explored this phenomenon during the early days of Nazi rule in 1934 in *The Mass Psychology of Fascism* and again in 1942 in *The Function of the Orgasm*. He argued that the First World War had destroyed many "compulsive authoritarian institutions" and had left the mass man faced with too much freedom. Centuries of slavery, oppression, and self-denial had generated a fear of freedom, and German Fascism had capitalized on "the conflict between longing for freedom and actual fear of freedom." The modern mass man fears political freedom as much as sexual freedom and longs for some powerful authoritarian figure to tell him what to do and reimpose the old restrictions under the guise of bringing liberation. This Hitler did exactly. He insisted that he was liberating the German people from their domination by Jews, Marxists, November criminals, pacifists, and the like, and at the same time imposed a rigid dictatorship. This particular combination of antithetical messages and conflicting actions responded to the antithetical, conflicting needs of the mass man. Hitler spoke of the evils of excessive sexual abstinence but diverted the focus of the sexual reform movement (in which Reich played a central role) from its purpose of restoring sexual pleasure to one of procreation. Hitler spoke of the sanctity of the German family but set up a series of institutions such as the Hitler Youth to take children away from parental rule, and he encouraged young men and women to produce children whether or not they were married. His promise to restore the "purity" of German blood was a disguised threat to take action against racial mixing. He exploited anti-Semitism effectively because behind the fear of Jewish money was a fear of Jewish sexuality. The "Jew" was a "filthy, sensual, brutally lustful slaughterer," and Hitler found a target for the fear of natural sexuality in the Jew. In Nazi Germany "the persecution of the Jews . . . stirred up the deepest antisexual defense functions of the antisexually brought up individual." Reich concluded his explanation of "fascist" sexual policy with a repetition of the underlying set of conflicting motives that governed the mass man. "Unconscious longing for sexual happiness and sexual purity, plus the simultaneous fear of normal sexuality and the abhorrence of perverse sexuality, results in fascist sadistic anti-Semitism." [38]

The sinister details of the Nazi regulation of sexuality are of great historical significance and must not be dismissed as accidental aberrations in modern history. Such regulation appealed to the fantasies of millions with its celebration of superior physical specimens, cleansed of disease and corruption and bred to physical perfection. That those goals were so naïvely and, ultimately, so perniciously conceived was tragic, but behind the racism and chauvinism was the vision of a population of healthy and beautiful bodies. Reich stood at the center of those conflicting ideals and watched the longing for physical excellence become spoiled by the clamor for racial purity and the suppression of the sensuous life-affirming impulses from which it had once arisen.

Chapter 18

The Philosophy of the Body

"Body am I, and soul"—thus speaks the child. And why should one not speak like children?

But the awakened and knowing say: body am I entirely, and nothing else; and soul is only a word for something about the body.

Nietzsche—*Thus Spoke Zarathustra*

In 1951 Maurice Merleau-Ponty speculated about the progress that the philosophical investigation of man had made during the previous fifty years. "Our century," he argued, "has wiped out *[effacé]* the dividing line between 'body' and 'mind,' and sees human life as through and through mental and corporeal, always based upon the body and always (even in its most carnal modes) interested in the relationships between persons." This historical claim is not entirely correct. The twentieth century did not "wipe out" the dividing line between mind and body, even though some thinkers tried. Merleau-Ponty was one of a number of philosophers who challenged the traditional dualism of mind and body and developed an approach to the study of man that sought to embrace both aspects of existence in a conceptual whole. In conclusion Merleau-Ponty added to his first bold claim a more modest appraisal: "The twentieth century has restored and deepened the notion of flesh, that is, of animate body." [1] This claim is correct. In this chapter I will discuss the nature of that restoration and deepening, and reconstruct in detail Sartre's contribution to it.

As early as 1781 Immanuel Kant argued in his *Critique of Pure Reason* that a duality between a mental and a physical substance is impossible, because "substance" is merely a category of experience, not an objective entity within the world. By destroying the validity of a substance metaphysics, Kant prepared the way for thinking about corporeal and mental existence in an essentially unified or at least an interrelated manner. In the late nineteenth century a number of thinkers began to focus on Kant's refutation of a dualism of substances and to explore the way a person experiences his body. In *Psychology From an Empirical Standpoint* (1874) Franz Brentano argued that every psychical phenomenon is "intentional." That is, every mental event, every act of perception is directed toward some object, and no mental event can take place without an object. In the representation something is always represented, in the judgment something is always judged, in desiring something is always desired. This theory influenced later studies of the experienced body and human relations. Sartre's argument that my experience of my body is always influenced by the way I experience the body of others and the way they experience mine derived from Brentano's theory of the intentionality of knowledge.

Henri Bergson supplied additional foundation for the investigation of the experienced body. In 1888 Bergson elucidated the contradictions inherent in traditional philosophy which attempted to explain the connection between non-spatial "mind" and spatially extended "body." [2] In *Matter and Memory* (1896) Bergson attempted to resolve this problem by arguing that the relation between mind and body could be solved by viewing all of matter as an ensemble of "images." "There is one such image which cuts across all others in that I know it not only from without by means of perceptions, but also from within by affections: this is my body." Much subsequent discussion of the way the body organizes the world around it was anticipated by Bergson's attempt to solve the traditional mind-body problem. His statement, "The objects which surround my body reflect the possible action of my body on them," anticipated the theory of perception developed by Merleau-Ponty fifty years later.[3]

Edmund Husserl developed the empirical method of Brentano into the phenomenological method of the twentieth century. Husserl sought to restore the validity of purely subjective knowledge by studying phenomena as they are experienced and by rigorously excluding from consideration "objective" theories about their causes or their internal nature. In 1913 Husserl made explicit the distinction between the body as experienced and the body as subject of scientific investigation with his use of the respective terms *"Leib"* and *"Körper"* for each.[4]

In 1914 the French philosopher Gabriel Marcel began to investigate the experienced body in his *Metaphysical Journal*. He sought a solution to the mystery of being, which he believed was linked with the relation of the body to consciousness. He wrote: "The problem of the reality of the body is shown to be the central problem, and upon its solution everything else depends."[5] Marcel rejected a number of alternative theories. He repeatedly denied that the body could be considered an instrument for perception or that sensations were messages from the body. Those theories presupposed the priority of mind over body. The body was not a thing that one could possess; rather, he argued, "I am my body." By 1921 he expressed the desire to obliterate all divisions between mind and body, self and world. "In other words," he wrote, "I want to show that I am really *attached* to and really adhere to all that exists—to the universe which is my universe and whose center is my body." And then a year later he continued: "The world only exists inasmuch as I can act on it: for there is only action inasmuch as I am my body. . . ."[6] Thus did Marcel endeavor to understand the mystery of his being with a philosophy that viewed his acting body as the source of all that exists.

The work which prepared the way for modern existentialist theories and profoundly influenced Sartre's philosophy of the body was Martin Heidegger's *Being and Time* (1927). Heidegger showed that one can inquire into the *meaning* of existence without first establishing which kinds of *things* make up the world (like "bodies" or "minds"). Thus I can ask directly how to think about my being without first having to ask what kind of thing I am. In Heidegger's view, my body is not then a special kind of thing but rather an essential part of my existence.[7]

To understand twentieth-century philosophy of the body it

is necessary to examine some of the ideas against which it was directed. The phenomenologists were particularly critical of Plato's theory of knowledge, Descartes's theory of human nature, and crude scientific materialism.

The phenomenologists objected to the Platonists' insistence that knowledge from sense experience is inferior to knowledge abstracted from experience. The phenomenologists held that knowledge acquired directly from the senses is valid. Moreover, they argued that it is not even meaningful to speak of senses as distinct from what they sense. They rejected the theory that the mind processes higher knowledge out of the raw material of "sense data," or that the body is an instrument for acquiring knowledge. The response of the body to the world is not some low-level preliminary act of perception to be further refined by higher mental activity, but rather that response is knowledge itself. They believed, in short, that perception, knowledge, and reality are one and the same, united in consciousness, which is the body.

The phenomenologists also criticized Cartesian dualism. Descartes theorized that mind and body are different substances, *res extensa* and *res cogitans*, that interact in the pineal gland. The phenomenologists avoided the difficulties inherent in the dualist position by rejecting the idea that mind and body have a separate existence or even a separate function. Descartes's famous proof of his existence, "I think, therefore I am," made no reference whatsoever to the existence of his body. The phenomenologists insisted that any proof of human existence would have to include the body. They also rejected theories that viewed the mind as residing *in* the body. They held rather that mind and body are two aspects of the same thing—existence. Although it was often difficult for them to give a clear positive formulation of their revision of traditional dualism, they persistently maintained the critical view that mind and body cannot be regarded as different substances or even as separate and autonomous concepts. One could not properly think about one without the other.

The phenomenologists repudiated the theories of those behaviorists who experimented on human beings by testing specific responses to stimuli. They objected that one cannot isolate a single stimulus on an individual and evaluate his response to that stimulus alone, because an individual always responds to the world as a whole and does so with his entire being. Therefore the controlled experiments of the behaviorists could not reveal anything valid about human nature. Furthermore, the behaviorists continued to assert the substantive theory of body, which the Kantian tradition had already rejected as impoverished.

The phenomenologists also rejected attempts to explain human existence in physio-chemical terms. Merleau-Ponty complained that "for many thinkers at the close of the nineteenth century, the body was a bit of matter, a network of mechanisms." He further charged that some naïve materialists "treated values, institutions, works of art, and words as a system of signs referring in the last analysis to the elementary needs of all organisms." One could not view man as a set of forces, isolated from the world as the scientists had done; rather one must try to understand man as he is in the world. Merleau-Ponty identified Freud as a key figure in this transition from "a conception of the body . . . which was initially that of a nineteenth-century doctor to the modern notion of the experienced body." Freud explained a child's development not merely in terms of bodily instincts, but as a pattern of exchanges between his body and that of his parents, a "system of attractions and tensions" with which he tries out various attitudes with his parents. "At least as much as [Freud] explains the psychological by the body, he shows the psychological meaning of the body, its hidden or latent logic. . . . With psychoanalysis, mind passes into body as, inversely, body passes into mind." To support his contention that the phenomenology of the body and psychoanalysis were part of a larger historical development, Merleau-Ponty discussed some imaginative writers who were also exploring the experienced body at that time. "With Proust, with Gide, an unwearying report on the body begins. It is confirmed, consulted, listened to like a person." For Proust the body became the "keeper of the past." It alone, despite the outer changes that transform it over time, is the sole bearer of our relationship between ourselves and our past.[8]

When Sartre began to write about the body, he had access to a variety of sources: the phenomenological method in Germany, Marcel's ruminations about the centrality of the body in human existence, Heidegger's distinction between entities and existence, Freud's interpretations of the meaning of the body, and a rich imaginative literature on the body incarnate. Sartre drew from these sources and integrated them into a comprehensive philosophy of the body that sought to change the way people thought about existence. He examined the way people talked about the body, the way they lived it. He stretched language to give precise descriptions of the experience of being incarnate flesh in a world of others.

In 1938 Sartre explored the nature of existence in an imaginative autobiographical journal, *Nausea*. The diarist Roquentin is plagued by nausea which he experiences at all times, but with particular intensity whenever he contemplates his body. "I see my hand spread out on the table. It lives—it is me." The hand takes on an identity of its own. "It is lying on its back. It shows me its fat belly. It looks like an animal upside down." The fingers become paws, the hand becomes like a crab, then like a fish. The weight of his arm pulls on him and reminds him of his entire body. "I can't suppress it, nor can I suppress the rest of my body, the sweaty warmth which spoils my shirt, nor all this warm obesity which turns lazily, as if someone were stirring it with a spoon." The experience of his body leads him to reflect on his existence. "I *exist*, I am the one who keeps it up. I. The body lives by itself once it has begun." The nausea accompanies this awareness of his body, and one day he discovers that the nausea is a key to existence. In a famous passage Roquentin records his insight upon looking at the root of a chestnut tree. "I understood that I had found the key to Existence, the key to my nausea, to my own life." Like the root of the tree, existence is rotten, bloated, obscene. It has no particular meaning or reason for being the way it is. "Every existing thing is born without reason, prolongs itself out of weakness and dies by chance." All is therefore superfluous and absurd. Roquentin finally concludes with a mixture of fascination and despair, "I *was* the root of the chestnut tree." Although the experience reveals to Roquentin the nature of all existence, there is particular similarity between his experience of the root and his body. The adjectives Sartre used to describe existence (the root)—flabby, languid, obscene, digesting, oily, callous, "like a bruise or a secretion, an oozing"—all suggest corporeality.[9] In his exhaustive study of the body in *Being and Nothingness* (1943), Sartre again introduced nausea as a basic feeling about our body and the world.

Being and Nothingness is a philosophical inquiry into the

nature of existence. Its subtitle, "An Essay on Phenomenological Ontology," would indicate that in it Sartre examines all of existence, but in truth the focus is on human existence. An important aspect of human existence is its incarnate nature, its rootedness in the body, and therefore a large portion of the work concerns the way we experience our body (which he called the body-for-itself) and the way we experience our body as known by others (which he called the body-for-others).

Sartre's account of the way we experience the body-for-itself included criticism of rival theories about the body and its relation to consciousness. He rejected the "objective" view of the body of science as a valid account of his body for himself. *My* stomach is not the same as the sack-shaped organ described in textbooks. I do not see my stomach and do not know for certain that it is there. I merely experience various sensations in my abdomen which I then attribute to the action of my stomach. A rigorous phenomenological account of my body, like one of my stomach, must exclude the objective body described by science and focus on the way the body is experienced, the body-for-itself. Sartre rejected the idea that the body is an instrument for receiving sensations. "It is the instrument which I cannot use in the way I use any other instrument." The camera remains an object even when it is recording, just as the eye of another person remains the same to me when it is perceiving. But *my* eye is different from either of these, because it ceases to be the same object for me when it functions. I see *through* my eye, and when I do, it is not an object. Therefore my experience of my eye differs fundamentally from that of all other eyes.

My body does not passively receive knowledge, because, as Sartre insisted, "sensation and action are rejoined and become one." Not only is my body actively engaged in the world; it is that by which there is a world. "To say that I have entered into the world . . . or that there is a world, or that I have a body is one and the same thing." And he continued enigmatically, "My body is co-extensive with the world, spread across all things, and at the same time it is condensed into this single point." [10] When he considered the way he felt his body, Sartre recalled his earlier reflections on nausea. Sometimes the body is a source of pleasure or of pain, but when there is no distinct pleasure or pain, consciousness continues "to have" a body. That constant awareness of the body, that "insipid taste which I cannot place," he wrote, "is what [I] have described elsewhere under the name of *Nausea*." This nausea is the foundation upon which our awareness of the body is based. "A dull and inescapable nausea perpetually reveals my body to my consciousness." [11] The experienced body is therefore different from the objective body described by science or discovered by observing the body of others. His inquiry was intended to elucidate that difference.

Following a short account of the way we experience the body of others, Sartre passed on to the more important part of his analysis—the way we exist as a body known by "the Other." In this section he developed his theory of human relations and established the basis for an existentialist ethic.

Sartre argued that the way one experiences one's body is affected by the presence of other people. "The nature of *our body for us* entirely escapes us to the extent that we can take upon it the Other's point of view." [12] For example, my knowledge of my stomach—that it is shaped like a bagpipe, that it produces enzymes, that it has an ulcer—comes from knowledge I have acquired from others. Therefore the way I experience my stomach is influenced by others, and if I allow their view of it to dominate, my subjective experience may escape me. I learn that my body is diseased by integrating my subjective experience of pain and illness with the scientific theory of a specific disease. "Pain" refers to specific sensations, "illness" refers to the recurrent pattern of pain in the same place at successive times, but "disease" refers to the construct of causes and symptoms explained to me by my physician.[13] Hence my knowledge of a disease is a mixture of my body as experienced and my body as known by others. When Sartre considered concrete relations with others, he explored this interrelation in detail.

Sartre believed that the essence of human existence is to be free. Human beings are condemned to choose what they will be, because they are nothing—they have no preordained essence. Most flee from that freedom and seek to embrace some system of thought to give their existence meaning and supply a set of rules to live by. But an individual can never escape the responsibility of his freedom. Efforts to flee always involve a self-deception that Sartre called "bad faith." Everybody is forever condemned to choose what they will be, and our freedom leads to an unceasing conflict with others.

This conflict is experienced in the corporeal sphere. Sartre explored that experience when he considered how the gaze of "the Other" affects the way I experience my body. "The Other's look fashions my body in its nakedness, causes it to be born, sculptures it, produces it as it *is,* sees it as I shall never see it." [14] The experience of shame and its corporeal manifestation in blushing illustrate the force of the look of another. Sartre offered the example of a man looking through a keyhole, absorbed in the goings-on in a room. Suddenly he hears footsteps. "Somebody's looking at me! What does this mean? It means that I am suddenly affected in my being and that essential modifications appear in my structure." [15] The mere awareness of the presence of another person transforms him from a man wholly absorbed in the scene through the keyhole into a man caught in the act of peeping. The Other has made him a voyeur, and he feels the gaze of the Other throughout his body. The ability of the other person to know me affects the way I experience my body and my shame. "The Other is the immense, invisible presence which supports this shame." But *my* look sets up the other term of the clash of perceiving bodies which is at the heart of concrete relations with others. "Thus in the look the death of my possibilities causes me to experience the Other's freedom." [16]

Sartre insisted that this awareness of the Other as manifested in shame is not something known, it is something lived. The problem of relations with others is not, as many philosophers had argued, a purely epistemological problem, because I am aware of their existence in action, and their presence is revealed to me throughout my body—in shame and in pride, in desire and in hate.

The conflict inherent in human relations is particularly vivid in love and sexual relations. Alone, the individual feels superfluous. In love, he feels that his existence is willed by another and is therefore justified. The Other's freedom to love gives his life meaning, and he therefore wants to possess that freedom. But if he succeeds in possessing the Other's freedom, the Other ceases to be free and loses the ability to give his life meaning. This conflict often leads to two basic responses—masochism and sadism.

The masochistic response involves "causing myself to be

absorbed by the Other and losing myself in his subjectivity in order to get rid of my own." [17] I renounce my freedom to avoid conflict with the Other. I offer myself as a passive object that the Other may debase at will. But the masochistic attitude ultimately fails. The Other will lose interest because I have renounced my freedom, or I will assert my freedom by having him torture me. I can never entirely annihilate my subjectivity, my power to judge the Other, nor can I entirely destroy his need for me as the passive object. Therefore masochism cannot resolve the conflict inherent in love. Sartre's argument that masochism is one of the few possible attitudes of human relations emphasized the sensuous underpinnings of human existence and suggested that human sexual relations were forever threatened with failure. The other possible response to these conflicts—sadism—further reveals the potential shipwreck of love.

Sartre wrote *Being and Nothingness* during the early years of World War II. He published it while the Nazis still occupied France and while Gestapo agents tortured prisoners to extract information. Sartre strongly identified with the Resistance, for which standing up under torture was the ultimate heroism. It was a time when human relations were strained by the exigencies of war, when heroes were reduced to cowards in the face of killing, and when the possibility of harmonious human relations appeared at an all-time low. It is not surprising that this study should have sounded a pessimistic note, and many critics have interpreted his final comment—"Man is a useless passion"—as his ultimate judgment on human existence. Sartre argued that he was defining the essence of human existence, valid for all places and times, but the traces of his personal experience are evident. His discussion of the conflict between people trying to maintain their freedom in a world of threatening "Others" echoed the hope of the French to recapture their political freedom. His dramas and short stories presented characters confronted by tormenting choices between life and death, torture and the betrayal of others. The extremes of cruelty and kindness were being enacted around him as he explored the nature of love, and although he insisted that the sadistic component was universal, there can be no doubt that some of the vividness of his treatment of it derived from historical circumstance.

Sartre prefaced his discussion of sadism with a definition of sexual desire and the caress. "My original attempt to get hold of the Other's free subjectivity through his objectivity-for-me is *sexual desire.*" He objected to the dismissal of sexuality as contingent and not essential to human existence. He explicitly rejected Heidegger's view of *Dasein* (human existence) as asexual and insisted that sexuality is a fundamental part of our relation with the world.[18]

Sexuality manifests itself first as desire, which reveals simultaneously one's own body and the body of the Other. It is not just a desire for contact with a purely material object, but with the Other's body in situation. For that reason a woman asleep cannot arouse the kind of desire that a conscious woman can. "A living body as an organic totality in situation with consciousness at the horizon: such is the object to which desire is addressed." Sexual desire involves a reciprocal sustaining of the meaning of two beings: ". . . when I grasp these shoulders, it can be said not only that my body is a means for touching the shoulders but that the Other's shoulders are a means for discovering my body as the fascinating revelation of facticity—that is, as flesh." [19] But that other body is never mere object, it is always a body in situation and therefore conscious. Through my relationship with that embodied consciousness my existence takes on added meaning.

The first concrete expression of desire is the caress, which Sartre defined as an appropriation of the Other's body. "The caress is not a simple stroking, it is a *shaping*. In caressing the Other I cause her flesh to be born beneath my caress, under my fingers." My caress is "the ensemble of those rituals which *incarnate* the Other . . . and cause the Other to be born as flesh for me and for herself." [20] But the caress also limits the freedom of the Other; it strips the body of its action by cutting it off from the possibilities which surround it. Here Sartre anticipated the ultimate restrictive gesture of sadism. The caress gives shape and meaning to the Other's body, but it also appropriates that body and robs it of its freedom by subjugating it to my desire. The ultimate goal of sexual desire involves a taking and a penetration. The caress wants "to impregnate the Other's body with consciousness and freedom." But if the Other's body is so impregnated with my freedom, it ceases to have the freedom that once rendered it desirable. The body that has been stripped of its freedom "falls from the level of *flesh* to the level of pure object." [21] This is the origin of sadism.

The sadist seeks to subjugate the free subject and render it an object. Whereas the initial inspiration of love, of desire, and of the caress is a reciprocity, sadism "wants the non-reciprocity of sexual relations." Whereas the loving caress seeks to give pleasure to the Other, the sadist seeks to destroy the freedom of the Other by inflicting pain. In contrast to the graceful movement of the free body, sadism likes to produce a special kind of incarnation called obscene. "The *obscene* appears when the body adopts postures which entirely strip it of its acts and which reveal the inertia of its flesh. The sight of a naked body from behind is not obscene. But certain involuntary waddlings of the rump are obscene." [22] Sartre explained this contrast from the fact that the flesh of the rump "cannot be justified by the situation." It does not contribute to the walking movement but rests uselessly on top of the legs. It is superfluous. It does not share in the movement of the body which in this instance is walking. The sadist seeks to destroy the grace and the freedom of the Other and render the entire body obscene. The sadist "wants to knead with his hands and bend under his wrists . . . the Other's freedom." The supreme moment for the sadist is when "the body is wholly flesh, panting and obscene." It is no longer free, but takes the form the torturers have given to it. Sartre focused on the moment of abjuration when the victim relinquishes all resistance and agrees to tell all or submit to the sadist's will. In that moment the freedom of the victim is suspended, but the sadist's pleasure is momentary and incomplete. Sadism does not solve the problem of love, because either the sadist becomes dependent upon his victim for the freedom that he seeks to possess or the victim affirms his freedom by choosing the moment of submission. Even the act of renouncing one's freedom requires a free choice, and therefore the goal of the sadist requires the cooperation of the Other. The sadist cannot resolve the conflict between free individuals that was the initial impulse for the sadistic attitude. Sartre did not argue that all love relations involve extremes of masochism or sadism, but he did insist that these attitudes are potential in all concrete relations with others. "Sadism and masochism," he concluded, "are the two reefs on which desire may founder." [23]

The thrust of Sartre's theory of human relations was to show that love involves the resolution of conflict between two free beings and that that resolution involves various modes of sexual desire.

Hence the body plays an essential role in human relations. Animate flesh determines the desire, the caress, and the attitudes of masochism, sadism, indifference, or hate that govern human relations. After Sartre the body could not be regarded as an accidental ingredient of human existence. It emerged in his philosophy as the foundation for the way we experience ourselves and relate to others. He sought to obliterate the dichotomy between mind and body, although one suspects that in spite of his search for terminology to break down the conceptual gap between the corporeal and the non-corporeal, the dualism remained as much a mystery to him as it had been to Freud or Marcel, or even to Descartes three hundred years earlier. Sartre's philosophy indeed broadened and deepened the notion of animate body, even if it did not entirely wipe out the dividing line between mind and body.

Perhaps Merleau-Ponty was not attempting to be literally correct when he argued that the twentieth century had wiped out the division. Perhaps he just wanted us to consider for a moment what such a conceptual revolution might be like.

Conclusion

Around the middle of the nineteenth century, Europeans and Americans began to reject the restrictive sexual morality that had come to govern their lives and adopt in its place one that gave a fuller understanding and enjoyment of bodily existence. I have led the reader through many phases of that cultural transformation and now will summarize its major aspects.

Advances in public health around mid-century profoundly influenced attitudes toward the body. The improvement of water supplies and sewage disposal helped make life in the cities cleaner and healthier and may well have affected attitudes toward body odor. The effort of some prominent intellectuals to catalog the multitude of smells that a human being could detect was symptomatic of a growing interest in expanding the range of human sensuous experience. Clothing reform did not make substantial headway until the end of the century, but the first important spokesmen made themselves heard around 1849, when Amelia Bloomer developed pants-like undergarments for women. Rubber condoms, which became available after 1843, offered protection against venereal disease, although Europeans did not use them in large quantities until after World War I.

Similar developments took place in high culture. In the early 1850s the Realists began to reject the idealized Classical nude and render the body as they found it in everyday life. The poetry of Baudelaire, Whitman, and Rossetti established a convention of explicit treatment of sexuality that challenged the more restrained conventions of the Romantic poets. Around mid-century, Marx began emphasizing the corporeal basis of history by insisting that bodily needs and laboring man must be the starting point of valid historical analysis. In a competitive society where profit and private ownership determine the exercise of human labor, all human relations—sexual and other—are alienated, and further, the labor process becomes so repetitious and meaningless that the laborer is alienated from his own body and is denied his potential for creative and ennobling labor. Darwin showed that the human body contained useless remnants of its animal ancestry and implicitly challenged the notion that man had been fashioned exactly in God's image. Privately Darwin argued the primacy of body over mind, and the broad cultural message of his theory was that the body plays a decisive role in the two basic goals of life: survival and reproduction.

In the 1860s a number of physicians, scientists, and popular philosophers speculated about the interaction of mind and body. Henry Maudsley began his pioneer work in psychosomatic medicine, and Thomas Huxley theorized how life might be explained in physio-chemical terms. Claude Bernard's studies of the action of the "interior milieu" of the body influenced the theories of temperament and heredity which later figured so prominently in the novels of Emile Zola. Psychologists worked to explain "higher" mental activity as a function of bodily processes. Studies of "physiological aesthetics" attempted to explain even the aesthetic response in purely organic terms, and Christian Science, which also dates from the 1860s, further emphasized the interaction of mind and body to a large popular audience.

In the 1870s medical literature on gender character began to look more favorably upon female sexuality. Marriage manuals showed a greater willingness to view the female orgasm as normal and insisted that traditional denunciations of female sexuality were misguided and destructive. At the same time the literature on male impotence reflected concern over the threatening potential of active female sexuality. This pairing of theories about male and female sexuality reveals a dialectical focal point of the transformation in sexual relations that was under way at that time. Medical literature also began to emphasize and exaggerate the bodily determination of hereditary transmission. Progress in the understanding of heredity proceeded at a rapid pace from the 1870s and led to a great deal of speculation about the extent to which parental traits were transmitted to children. Confusion about the respective roles of each of the three possible paths of communication—hereditary, infectuous, and cultural—led many Victorians to believe that children were fated inevitably to inherit the bodily features, temperaments, germs, and ideas of their parents.

The reader must be wondering where, amidst all of these trends, are the major turning points and "events" in the history of the body. Of course these changes did not occur quickly, but beginning in the 1880s there was a striking rise in the volume and quality of the literature on sexuality. In addition to the explicit treatment in the naturalistic novel and the art of the period, a number of scientific studies analyzed the components of the sexual impulse and their maturation during development. In the 1880s psychiatrists elaborated on Charcot's charts of the hysterogenic zones and mapped out a number of erotogenic zones of the body in addition to the genitals. They concluded that almost any surface of the body could generate sexual excitation. The publication of Krafft-Ebing's *Psychopathia Sexualis* in 1886 marks the beginning of the contribution of formal scholarship to the modern sexual revolution. From that time researchers and popular writers compiled tomes of information from psychiatric, anthropological, and historical studies of sexuality. At first all deviations from "normal" sexuality were viewed as pathological, but by the turn of the century the medical literature began to suggest that variations on the traditional routine might help the achievement of full sexual satisfaction. In the 1890s the rigid boundaries between normal and pathological sexuality that Krafft-Ebing had established began to erode. The trial of Oscar Wilde in 1895 publicized the dilemma confronting a homosexual in an age of narrowly defined sex roles. Although his personal struggle with that morality ended in tragedy, it led to a reconsideration of the rationale for a sexual ethic that could destroy a leading cultural figure. The Wilde case inspired a number of publications in the first decade of the century that argued the universality of bisexual impulses and challenged traditional sex roles.

The turn of the century witnessed the rise of a physical culture movement all over Europe as thousands became persuaded that hiking and fresh air, exercise and bicycling would improve their health. The clothing reform movement achieved widespread support at that time as men and women joined in the protest against tight lacing. Programs of exercise for men and women were devised to lessen the physical and emotional problems attributed to the stress of urban life: nervous tension, physical weakness, and even psychic impotence. Around 1900 Isadora Duncan explained the insight that led to her innovation in modern dance: that the source of all movement was located in the solar plexus and not at the base of the spine as classical ballet had taught. Her "liberated" movements became a symbol of the liberation of the body sought by the champions of physical culture and sex reform.

Freud developed psychoanalytic theory in the 1890s, and it began to attract a small group of followers after the turn of the century. Psychoanalysis offered a detailed explanation of the ways the body influenced character and marked an enormous contribution to the discovery of the meaning of the body. As Merleau-Ponty remarked, Freud revealed the "latent logic" hidden in the body. In the years immediately preceding the war the Expressionists depicted nudes in wildly exaggerated shapes and positions. The passion that had been suppressed in nineteenth-century art emerged in their brightly colored paintings as a challenge to the sexual morality that they believed had so perniciously restricted human feeling and bodily movement.

Thus by the eve of World War I, Europe was well prepared in theory for a revision of its sexual morality. The war made many of these ideas a reality. It radically transformed the way people dressed, understood, enjoyed, and cared for their bodies, and it forced Europeans to give up the elaborate decorum that had ruled since the Victorian era. The disruption of traditional patterns of male-female relations lasted for four years and led to experimentation with new sex roles. Women ventured out of the home and took the opportunity to look after their own sexual needs, while the men in the trenches were forced to do without sex or to make do with masturbation, prostitution, homosexuality, or even sodomy. The filth, the disease, the brutality, the pain, and the killing further affected the way men experienced their bodies. Observers reported that both sexes walked differently after the war, and when the men returned it was impossible to resume the prewar sex roles. When Johnny came marching home again, his women hardly knew him.

The cultural record of the postwar years illustrates the extent of the changes that had taken place. The fiction of D. H. Lawrence and Henry Miller is a joyous celebration of the emancipation of the flesh. Lawrence described the bodies of Connie and Mellors in *Lady Chatterley's Lover* as if he were giving Europeans their first lesson in anatomy. He offered a graphic picture of a sexually liberated couple making love in the woods, unmindful of the protests of Connie's anti-sexual husband, crippled by the war, imprisoned in his wheel chair, futilely espousing the primacy of mind over body. Lawrence was merciless in his attack as he portrayed prewar sexuality resting limp and powerless in the loins of the invalid aristocrat, Lord Clifford Chatterley. Miller appreciated the destructive potential of the old morality, but in contrast to the solemnity of Lawrence he scoffed at those who remained under its sway.

Not all efforts at the liberation of the body had a positive effect. The physical culture movement in Germany shows some problems that can arise when a state undertakes to promote the racial uniformity and bodily excellence of its population. The movement reveals both the life-affirming inspiration of European physical culture generally and the life-denying values of the chauvinism that so often accompanied it. In Nazi Germany the authoritarian tradition filtered down to the psychosexual level and generated intolerance for bodily types that did not conform to the officially sanctioned racial ideal. In the later years of the war this attitude developed into an extreme fanaticism that proclaimed the need to cultivate "good" German bodies and destroy "bad" foreign bodies. Hitler's formula for racial excellence—"that the most beautiful bodies should find one another"—neatly summarized his intention to breed an elite of superb physical specimens.

Although the philosophy of the body should not be regarded as a direct consequence of World War I, it constitutes part of a broad cultural reorientation about the body that emerged in the postwar years. After struggling with Cartesian dualism for over two centuries, twentieth-century phenomenologists scrapped the search for *what* we are and concerned themselves with the *way* we are. Sartre explored existence as he found it, an embodied consciousness acting in a world of other similarly constituted beings, forever confronting dissolution into nothingness. The body is neither a stable center of our being nor a sanctuary from the stickiness of our relations with others. As European culture emerged from its long devaluation of the body as the "lower" half of being, Sartre insisted that the body was not inferior, rather it was the very same as our being. But far from exalting the body, Sartre portrayed it as a slippery, honeylike amorphousness that we must endow with shape (i.e., meaning) at every moment. He viewed sexual relations as a nightmarish struggle to avoid sadistic or masochistic attitudes. Yet Sartre also embraced a theory of alienation, which he formulated in terms of the loneliness and abandonment of our separate existences. He thus left his readers with a paradox: individuals are threatened by both alienation from, and enslavement to, one another.

The modern age is still trying to find a "healthy" sexual morality. As long as physical beauty determines sexual choices, human relations will be guided by fortuitous, pleasing compositions of bone, muscle, and skin. Elites of the beautiful will continue to live privileged lives, and character and "inner" beauty will continue to take second place in the contests for sexual partners. It is tempting to condemn this rewarding of physical beauty by pointing to the riches and fame we shower on fashion models and athletes while the aged and the deformed are hidden away in poverty and neglect. But perhaps our condemnations are in bad faith and even a bit silly. Vanity and the appreciation of bodily excellence occur in all cultures, and they are the source of those energies that make possible the continuation of life. It is, after all, the physical expression of sexual desire that produces offspring. Though erotic stimuli vary from one age to another, they are always based on physical attraction, and as long as that is the case, we will remain dominated by the mysterious and enduring power of the body to generate sexual tension and release sexual pleasure.

In spite of the efforts of recent years, sexuality remains a source of conflict. The permissive sexual morality of the 1960s did not eliminate all sexual frustration, and in some ways the preoccupation with sexual performance itself became a problem. Change in the direction of greater equality may in the long run create more gratifying sex roles, but during our period of transition there has occurred a rise in anxiety about male impotence similar to the anxiety of the late Victorian period when feminists began to challenge the convention that women remain sexually passive and show no pleasure. The body has not been able to keep pace with the radical demands of the Women's Liberation Movement, even among those committed to it. Sexual mores die hard, and the hopeful expectation that male-female relations would conform to the demand for sexual equality contrasts with the conservative body lag that has tended to maintain traditional roles requiring active men and passive women. Someday, when men and women come to have equal work responsibility and social status, we may achieve a more varied and more imaginative distribution of sex roles, and perhaps then we will be able to improve upon the current nature/nurture debate and ascertain more accurately to what extent sex roles are determined by anatomy.

These critical considerations, however, need not discourage the search for a sensible sexual morality. The progress made over the past century and a quarter has been impressive. The accumulation of knowledge and its presentation to the public has broken the conspiracy of silence that so paralyzed the Victorians. Reformers have used that information to create a sex ethic to help people regulate their lives in conformity with bodily needs rather than restrictive social requirements. Psychiatrists have analyzed the problem of excessive sexual repression, and imaginative writers have dramatized its destructive effect on the individual. The historical record shows that although the problems of the body are universal, the ways we deal with them are subject to change. The body is a great source of pain, but it can be equally generous in its offerings of pleasure. Perhaps through rational inquiry and bold experimentation we can learn to reduce the unnecessary suffering of existence and give ourselves a fuller share of the pleasure.

Author's Note

Some of the research for this book was made possible by a grant from the National Institute of Mental Health which enabled me to participate in certain aspects of a psychiatric residency program at Cornell Medical College from 1966 to 1970 while I was also writing my dissertation in history at Columbia University. During that time I attended a study group in the history of psychiatry under the direction of Eric T. Carlson which provided an opportunity to exchange ideas with historians and psychiatrists who shared common interests. Since then the History Department at Northern Illinois University has provided continuing support for my research. I would like to thank Samuel Huang and Myrtie Podschwit of the Interlibrary Loan Department of Northern Illinois University for providing so many hard-to-find sources. I am also grateful to Darla Woodward and the departmental secretaries who typed the manuscript with such good cheer. My subject has engaged the interest of almost everybody I have spoken to over the years, and I have had most helpful discussions about it with William Beik, Lloyd DeMause, Emory G. Evans, Mary Furner, Harvey Goldberg, James Greenlee, J. Carroll Moody, Heinz Osterle, Marvin Powell, Richard Price, Albert Resis, Marvin Rosen, and J. Harvey Smith. I would also like to thank Joan Severa and Edward Maeder, who introduced me to the history of fashion and guided me about the collection of costumes of the Wisconsin State Historical Society. Dorathea Beard, Henry Ebel, and Jerry Meyer offered some guidance in the history of art, and Michael Gelven, Nancy Metzel, Sherman Stanage, and Kenneth Winston

AUTHOR'S NOTE

supplied valuable comments on my treatment of philosophical works. I am particularly grateful to those who read and commented on parts of the manuscript in its final stages: Jo Ann Arpin, Stephen Foster, Charles H. George, Margaret George, John S. Haller, Nao Hauser, Benjamin Keen, William Logue, and George Mosse. Special thanks are due to Martin J. Sklar for an endless source of good sense, good humor, and expert criticism.

My greatest debt is to Rudolph Binion for the inspiration and guidance he has so generously given over the years.

Dekalb, Illinois, 1974

NOTES

CHAPTER 1

1. I have used "Victorian" to refer to the latter two-thirds of the nineteenth century in Europe, not just in England, and "Victorianism" to refer to the strict sexual morality associated with European society at that time.

2. Quoted in Maurice J. Quinlan, *Victorian Prelude: A History of English Manners*, New York, 1941, p. 1. Another example of this shift is reported by Peter T. Cominos, "Late-Victorian Sexual Respectability and the Social System," *International Review of Social History*, VIII, 1963, p. 225: "No better proof of the changes that had come over English morals between 1765 and 1825 could be found than the difference in public attitude toward [Dashwood] . . . and Byron after their scandals became known. The sexual excesses and sacrilegious impieties of Sir Francis in the 1760s approximately balance Lord Byron's relations with Mrs. Leigh and the scandals of his own affairs, yet Byron was frozen into exile, while Dashwood became Postmaster General of Great Britain and remained the friend of England's great."

3. Milton Rugoff, *Prudery and Passion*, New York, 1971, p. 108. Rugoff adds a quotation from a doctor who in 1852 commended this prudery. "I am proud to say that in this country generally . . . women prefer to suffer the extremity of danger and pain rather than waive those scruples of delicacy which prevent their maladies from being fully explored."

4. D. Mensinga, *Wider die Verunstaltung und Schädigung des weiblichen Körpers*, Berlin, 1898.

5. Richard Lewinsohn, *A History of Sexual Customs*, New York, 1958, p. 268.

6. F. Marryat, *Diary in America*, Philadelphia, 1839, Vol. 2, p. 45.

7. Martin Freud, *Sigmund Freud: Man and Father*, New York, 1958, p. 60. The German gynecologist C. H. Stratz in *Die Frauenkleidung*, 1900, added another anecdote concerning a little girl who remarked upon seeing her aunt in a cycling costume: "Mamma, die Tant hat ja Beine," p. 389.

8. Carl H. Degler, "What Ought to Be and What Was: Women's Sexuality in the Nineteenth Century," *The American Historical Review*, December, 1974, discusses a number of sources from the 1870s and 1880s that refer to female orgasms, although he does mention one reference as early as 1866, p. 1475.

9. Quoted in E. P. Thompson, *The Making of the English Working Class*, New York, 1963, p. 402.

10. Harold Nicolson, *Good Behavior*, New York, 1956, p. 235. Richard Soloway, *The Onslaught of Respectability—A Study of English Moral Thought During the French Revolution: 1798–1802*, University of Wisconsin Dissertation, 1960, argued, following Nicolson, that fear of the lower classes led to a tightening of the morality of the middle and upper classes, especially after 1792–1793 and the execution of the king. One example he reported was that taking snuff was still fashionable in 1786, but "was clearly considered indelicate by the late 1790's," p. 178.

11. Wilhelm Reich, "Dialectical Materialism and Psychoanalysis," in *Sex-Pol Essays 1929–1934*, ed. Lee Baxandall, New York, 1966, pp. 49–50.

12. Quinlan, *Victorian Prelude*, p. 178.

13. Steven Marcus, *The Other Victorians*, New York, 1964, p. 146.

14. Jos van Ussel, *Geschichte der Sexualfeindschaft*, Hamburg, 1970, pp. 15, 39, 56.

15. Quinlan, *Victorian Prelude*, p. 143, concluded that "the educators tabooed the freedom for which Mary Wollstonecraft had pleaded. Instead of lowering the barrier between the sexes, they insisted upon raising it still higher." Soloway, *Onslaught of Respectability*, elaborated upon the political motives behind this reassertion of male domination. "The chaste, undefiled woman was a symbol of the stability and honor, and . . . very often represented the national salvation of England," p. 290.

16. John S. Haller and Robin M. Haller, *The Physician and Sexuality in Victorian America*, Chicago, 1974, pp. xiii–xiv.

17. Cominos, "Late Victorian," p. 37.

18. Lewis Mumford, *Technics and Civilization*, New York, 1934, p. 180.

19. Thompson, *Making of English Working Class*, pp. 370–372.

CHAPTER 2

1. My observations on the history of fashion are drawn largely from James Laver, *The Concise History of Costume and Fashion*, undated; C. Willett and Phyllis Cunnington, *The History of Underclothes*, London, 1951; R. Broby-Johansen, *Body and Clothes*, 1968; Ellen Moers, *The Dandy: Brummel to Beerbohm*, 1960; Cecil Saint-Laurent, *A History of Ladies' Underwear*, 1966; Norah Waugh, *Cut of Women's Clothes 1600–1930*, New York, 1968.

2. Laver, *Concise History*, p. 168.

3. Peter Quennell, *Victorian Panorama*, London, 1937, p. 92, commented on the significance of the crinoline: "Henceforth woman was not a two-legged viviparous animal, but an exquisite and unreal being who moved without any apparent means of locomotion, in a perpetual sighing rustle of silken drapery." Bernard Rudolfsky, *The Unfashionable Human Body*, New York, 1971, offered a similar interpretation of the monobosom fashion of the late nineteenth century as a sartorial disguise of the fact that women have two breasts—a biological fact that reminded man of his animal origins, p. 44.

4. Quennell, *Victorian Panorama*, p. 96.

5. J. C. Flugel's psychoanalytically oriented study, *The Psychology of Clothes*, New York, 1930, commented on this shift: "The Great Masculine Renunciation . . . took place at the end of the eighteenth century when man abandoned his claim to being considered beautiful. He henceforth aimed at being only useful," p. 112.

6. Thomas Carlyle, *Sartor Resartus: The Life and Opinions of Herr Teufelsdröckh*, London, 1836.

7. Charles Baudelaire, "Le Dandy," in *Oeuvres Complètes de Baudelaire*, Paris, 1961, p. 1179.

8. Max Beerbohm, "Dandies and Dandies," in *Works*, 1896. The debate continues. Ellen Moers concluded that the dandy was basically "antibourgeois" with a strong nostalgia for the lost aristocracy, while James

Laver insisted that "the revolution [the dandy] symbolized was essentially a conspiracy against the aristocracy." Ellen Moers, *The Dandy*, p. 14; Laver, *Concise History*, p. 34.

9. For every advocate of reform there was someone else to defend the *status quo*. The German critic Karl Erdmann, for example, criticized those who wanted to introduce "unmanly" colors into men's fashion. He further complained that previously one could tell rank and social position by clothing, but that was no longer the case in a "democratic" society where "the executioner and the baron wear the same clothes." *Alltäglichen und Neues*, Leipzig, 1898, p. 129.

10. Quoted by Rudolfsky, *Unfashionable*, p. 163.

11. Roxey A. Caplin, *Health and Beauty; or, Corsets and Clothing, constructed in accordance with the physiological laws of the Human Body*, London, 1856.

12. Ada S. Ballin, *The Science of Dress in Theory and Practice*, London, 1885, p. 160.

13. Emanuel Hermann, *Naturgeschichte der Kleidung*, Vienna, 1878, p. 261.

14. Paul Schultze-Naumberg, *Die Kultur des weiblichen Körpers als Grundlage der Frauenkleidung*, Jena, 1901. See especially pp. 135–141.

15. Christian H. Stratz, *Die Frauen auf Java: Eine gynäkologische Studie*, Stuttgart, 1897.

16. Christian H. Stratz, *Die Frauenkleidung und ihre natürliche Entwicklung*, Stuttgart, 1900, p. 300. Stratz explained some of the popular slang for various states of corporeal decline caused by corsets. The fold over the navel was called a *"Spitzbauch"* (pointed belly); the flesh that would hang out when the corset was removed was a *"Hangebauch"* (sagging belly); and in extremely corpulent women it was called a *"Froschbauch"* (frog belly).

17. Felix Dietrich ed., *Neue Männer-Kleidung*, Leipzig, 1912, pp. 15–16.

18. Arthur Marwick, *The Deluge*, New York, 1970, p. 111.

19. *Ibid.*, p. 112.

20. Saint-Laurent, *History of Ladies' Underwear*, p. 157.

21. Hermann Broch, *The Sleepwalkers* (1931), translated from the German, New York, 1964, p. 21.

22. Broby-Johansen, *Body and Clothes*, p. 221.

23. Carlyle, *Sartor Resartus*, p. 3.

24. Hermann Lotze, *Microcosmus*, 1856–1864, quoted by Flugel, *Psychology of Clothes*, pp. 592–595.

25. Thorstein Veblen, *The Portable Veblen*, New York, 1948, pp. 199–200.

26. Eduard Fuchs, *Illustrierte Sittengeschichte vom Mittelalter bis zur Gegenwart*, Vol. 3, p. 214.

27. Flugel, *Psychology of Clothes*, p. 45.

CHAPTER 3

1. I have used the following histories of the nude in art. Christian H. Stratz, *Die Körperformen in Kunst und Leben der Japaner*, Stuttgart, 1902, argues that the Japanese have a "highly developed feeling for natural beauty as reflected in their art" in contrast with Europeans, who are smothered by artificial conceptions of beauty. Joseph Kirchner, *Die Darstellung des ersten Menschenpaares in der bildenden Kunst*, Stuttgart, 1903, surveys canvases depicting Adam and Eve and lacks any historical explanation of changing views of the pair. Julius Lange, *Die menschliche Gestalt in der Geschichte der Kunst*, Strassburg, 1903, traces a general development from the Greek concern with the "purely spatial and outer side" of the human form to an interest in the subjective life as reflected in the body. Jean Cassou, *The Female Form in Painting*, London, 1953, offers a brief survey of the changing conception of the body as reflected in Greek, Christian, Renaissance, and Enlightenment art, and then supplies a short commentary on a number of nudes from the nineteenth and twentieth centuries. Kenneth Clark, *The Nude: A Study in Ideal Form*, Washington, 1956, is still the most useful general history of the subject. He is a bit rough on Wilhelm Hausenstein's *Der nackte Mensch*, 1913, which he evaluates as "much useful material cooked into a Marxist stew," p. 5. Joseph Relouge, *Meisterwerke der Aktmalerei*, Gütersloh, 1966, is useful largely for the reproductions of nineteenth-century nudes. Edward Lucie-Smith, *Eroticism in Western Art*, New York, 1972, offers a good introduction to erotic art. Linda Nochlin ed., *Woman as Sex Object*, New York, 1972, is a collection of articles on the rendering of women in modern European art.

2. John L. Connolly, Jr., "Ingres as the Erotic Intellect," in *Woman as Sex Object*, Linda Nochlin ed., New York, 1972, pp. 17–31.

3. Clark, *The Nude*, p. 443, concludes that "Courbet, for all his defiant trumpetings, continued to see the female body through the memories of the antique."

4. The artistic community still has not resolved this problem, although

most artists would agree that art ought not to produce physical excitation in its audience.

5. Beatrice Farwell, "Courbet's *'Baigneuses'* and the Rhetorical Feminine Image," in Nochlin, *Woman as Sex Object*, p. 75. On the general problem of the artist's struggle to avoid painting pubic hair see also Ronald Pearsall, *The Worm in the Bud*, London, 1969, pp. 142–146. Pearsall observes that William Etty, who painted very popular nudes in the Victorian period, "conformed to the unwritten code that specified no pubic hair." The adolescent girl became a popular subject for nude studies "if only for the absence of voluptuousness and pubic hair. . . . A thesis could be written on the effect of pubic hair on Victorian sexual thinking. Pubic hair was the omnipresent reminder of the animal in man, the hairy beast brought to the knowledge of the shocked middle classes by Darwin."

6. Victor Hugo, "Preface to Cromwell" (1827), in Eugen Weber ed., *Paths to the Present*, New York, 1965, p. 42.

7. Karl Rosenkranz, *Aesthetik des Hässlichen*, Königsberg, 1853.

8. Charles Baudelaire, *Oeuvres Complètes de Baudelaire*, Paris, 1961, p. 23.

9. *Ibid.*, pp. 29–31.

10. The critic Geoffrey H. Hartmann, *The Unmediated Vision: An Interpretation of Wordsworth, Hopkins, Rilke, and Valéry*, 1954, has interpreted the work of his four titular figures as an effort to break with the Judeo-Christian view of the incorporeality of man, which for Rilke in particular involved the exploration of the morbid and the ugly. Rilke's *The Notebooks of Malte Laurids Brigge* (1910) includes some vignettes of dying, and, Hartmann argued, "would have been impossible without Baudelaire and his poem 'Une Charogne' or Flaubert and 'Saint-Julien l'hospitalier.' These works of art show that the aesthetic contemplation overcame itself to such an extent that it was able to discover and accept the essential reality of the horrible and the repulsive," p. 135.

11. Clark, *The Nude*, p. 445.

12. Lucie-Smith, *Eroticism*, wrote: "The gleeful, rather frivolous celebration of the pleasures of the sexual act, such as we find in a picture like Fragonard's *The Happy Lovers*, gives place to the embodiment of feelings of doubt and anguish in the prints of leading Expressionists such as Kirchner and Heckel," p. 135.

13. Frank Whitford, *Expressionism*, London, 1970, p. 64.

14. Lucie-Smith, *Eroticism*. "Like Beardsley, Klimt was fascinated by the

more perverse aspects of human sexuality, and even when he is content to show us a pair of lovers embracing, the cramped attitude of the figures somehow suggests that there is something unwholesome or even demoniacal about their feeling for each other," p. 145.

15. Gerald Needham, "Manet, 'Olympia' and Pornographic Photography," in Nochlin, *Woman as Sex Object*, pp. 81–83.

16. Béla Bélazs, *Der sichtbare Mensch: Eine Film-Dramaturgie*, Halle, 1923, pp. 24–30.

CHAPTER 4

1. William Paul Gerhard, *Half-century of Sanitation: 1850–1899*, New York, 1899, p. 25.

2. René Dubos, *Louis Pasteur: Free Lance of Science*, New York, 1950, p. 295.

3. Guy Thuillier, "Pour une Histoire de l'hygiène corporelle," *Revue d'histoire économique et sociale*, Vol. XLVI, 1968, pp. 232–253.

4. Quoted in Eugen Weber, *A Modern History of Europe*, New York, 1971, p. 994.

5. *Bilder-Lexicon Kulturgeschichte*, published by *Institut für Sexualforschung* in Vienna, Vol. I, 1928, p. 541.

6. Edwin Chadwick, *Report on the Sanitary Condition of the Labouring Population of Great Britain, 1842*, Edinburgh, 1965, p. 423.

7. Dubos, *Louis Pasteur*, p. 314.

8. Dubos concluded that "by 1875, the association of microorganisms with disease had received fairly wide acceptance in the medical world." *Ibid.*, p. 247.

9. William James, *The Principles of Psychology*, 1890, Vol. II, pp. 437–438. An interesting contrast to this speculation is the fact that a full-length study by Dr. E. Monin, *L'Hygiène des riches*, 1891, surveyed a number of diseases to which the rich were believed to be especially susceptible such as migraines, ulcers, gout, arthritis, asthma, etc., but made no reference to any diseases that the rich feared that they were likely to contract from contact with the poorer classes.

10. Sylvanus Stall, *What a Young Husband Ought to Know*, London, 1897, p. 110.

11. Thomas Mann, *The Magic Mountain*, translated from the German by H. T. Lowe-Porter, New York, 1966, pp. 250–251.

12. *Ibid.*, p. 610.

13. *Ibid.*, pp. 342–343.

14. Paul Ricord mentions these in an article on mercury treatment that appeared in *The Lancet*, 1848, pp. 543–544.

15. William Acton, *A Practical Treatise on Diseases of the Urinary and Generative Organs*, London, 1857, p. 439.

16. Ronald Pearsall, *The Worm in the Bud*, London, 1969, p. 287. See also Irene Clephane, *Towards Sex Freedom*, London, 1935, on the Royal Commission on Venereal Diseases (1914–1916) which described the "abominable superstition" that a man can rid himself of venereal disease by intercourse with a virgin, p. 44.

CHAPTER 5

1. In *Studies in the Psychology of Sex: Sexual Selection in Man*, Philadelphia, 1905, p. 79, Havelock Ellis concluded: "Most of the writers on the psychology of love at this period, however, seem to have passed over the olfactory element in sexual attraction, regarding it probably as too unaesthetic. It receives no emphasis either in Senancour's *De l'Amour* (1808), Stendhal's *De l'Amour* (1822), or Michelet's *L'Amour* (1858)."

2. Terence McLaughlin, *Dirt: A Social History as Seen Through the Uses and Abuses of Dirt*, New York, 1971, p. 148.

3. Quoted in René Dubos, *Louis Pasteur: Free Lance of Science*, New York, 1950, p. 305.

4. Compare the situation in the late nineteenth century with that of the seventeenth as recounted in a recent history: "In the neighborhood of the Louvre, in several parts of the court, on the great stairway, and in the passageways, behind the doors, and just about everywhere, one sees a thousand ordures, one smells a thousand intolerable stenches, caused by the natural necessities which everyone performs there every day." David Hunt, *Parents and Children in History*, New York, 1970, p. 141.

5. Ellis, *Studies*, p. 78.

6. Some early pioneers of the study of sex who accepted Haeckel's theory were Charles Féré, Wilhelm Bölsche, and Iwan Bloch.

7. Wilhelm Fliess, *Die Beziehung zwischen Nase und weiblichen Geschlechtsorganen*, Leipzig, 1897.

8. Albert Hagen (pseud. for Iwan Bloch), *Die sexuelle Osphresiologie: Die Beziehung des Geruchsinnes und der Gerüche zur menschlichen Geschlechtstätigkeit*, Breslau, 1900.

9. J. H. Colhausen, *Von der seltenen Art, sein Leben durch das Anhauchen junger Mädchen bis auf 115 Jahr zu verlängern*, Stuttgart, 1847.

10. Ellis, *Studies*, p. 110.

11. Sigmund Freud, *Civilization and Its Discontents*, New York, 1961, p. 53.

12. Gustav Jaeger, *Die Entstehung der Seele*, Leipzig, 1881.

13. Gustav Jaeger, *Health-Culture and the Sanitary Woolen System*, translated from the German, Chicago, 1886, p. 45.

14. Ernest Monin, *Les Odeurs du corps humain*, Paris, 1903.

15. Carl Giessler, *Wegweiser zu einer Psychologie des Geruchs*, Hamburg, 1894.

16. Heinrich Pudor, *Hohe Schule des Sinnenlebens*, Munich, 1895.

17. Cited in Enid Starkie, *Baudelaire*, London, 1957, p. 272.

18. Charles Baudelaire, "Correspondences," translated by Eugen Weber, *Paths to the Present*, New York, 1965, p. 204.

19. Charles Baudelaire, *Oeuvres Complètes de Baudelaire*, Paris, 1961, p. 25.

20. *Ibid.*, pp. 252–253.

21. Léopold Bernard, *Les Odeurs dans les romans de Zola*, Montpellier, 1889, p. 11.

22. Emile Zola, *Les Rougon-Macquart*, Vol. II, Paris, 1961, p. 1206.

23. Cited in Ellis, *Studies*, p. 81.

CHAPTER 6

1. Robert C. Tucker, *The Marx-Engels Reader*, New York, 1972, pp. 119–120.

2. Friedrich Engels, *Ludwig Feuerbach and the Outcome of Classical German Philosophy*, New York, 1941, p. 18.

3. Ludwig Feuerbach, *The Fiery Brook: Selected Writings of Ludwig Feuerbach*, ed. by Zawar Hanfi, New York, 1972, p. 227. Karl Löwith, *From Hegel to Nietzsche*, New York, 1961, argued the sexual nature of the sensuousness which Feuerbach believed dominated the world, p. 77.

4. Lewis Feuer pointed out the specifically sexual striving of the Feuerbachians in "What is Alienation? The Career of a Concept" (1962), reprinted in *Marx and the Intellectuals*, New York, 1969: "The concept 'alienation' expressed the striving of the romantic movement, the recovery of spontaneous emotional life. There was a revulsion against sexual asceticism, a rediscovery among the German intellectuals of physical pleasure. German philosophy, the product of theological seminaries, had negated the human body; the new philosophers, disciples of Ludwig Feuerbach, affirmed it. The root meaning of 'alienation' for Feuerbach was, it must be emphasized, sexual; the alienated man was one who had acquired a horror of his sexual life . . . ," pp. 73-74.

5. Karl Marx, *Early Writings*, translated by T. B. Bottomore, New York, 1963, p. 127.

6. *Ibid.*, p. 159.

7. Tucker, *Marx-Engels Reader*, p. 93.

8. Marx, *Early Writings*, p. 154.

9. *Ibid.*, p. 171.

10. Tucker, *Marx-Engels Reader*, p. 113.

11. *Ibid.*, p. 118.

12. Karl Marx, *Capital*, Vol. I, New York, 1906, pp. 197-198.

13. Thomas Hanna came to a similar conclusion in *Bodies in Revolt*, New York, 1970, which surveyed a number of European intellectuals' views of the body. Hanna concluded that Marx, whom he described as "a somatic thinker par excellence, . . . was the first man to see penetratingly into the primitive biological processes which underlay not simply 'economic' but all of human history," pp. 164, 172. Maurice Merleau-Ponty interpreted Marxism as having shown that "there is a flesh of history in which (as in our own body) everything counts and has a bearing." *Signs*, translated from the French, Northwestern University Press, 1964, p. 20.

14. Charles Darwin, *The Origin of Species and the Descent of Man*, New York, Modern Library, p. 911. "Arboreal" means "living in or among trees."

15. *Ibid.*, pp. 395-398.

16. *Ibid.*, p. 442.

17. *Ibid.*, p. 446.

18. *Ibid.*, p. 892.

270   ANATOMY AND DESTINY

19. Quoted in Howard E. Gruber, *Darwin on Man: A Psychological Study of Scientific Creativity*, New York, 1974, p. 217.

20. Ernest Becker, *The Revolution in Psychiatry*, New York, 1964, p. 17.

CHAPTER 7

1. Julien Offray de Lamettrie, *Man a Machine*, Open Court, 1961, p. 148.

2. Arthur Schopenhauer, *The World as Will and Representation* (1818), Clinton, Massachusetts, 1958. In a supplement of 1844 Schopenhauer added a bit of playful mockery of human love. "Why all this noise and fuss? Why all the urgency, uproar, anguish and exertion?" he asked. "It is merely a question of every Jack finding his Jill." In a footnote he explained that he dared not express himself precisely in that passage and requested that "the patient and gracious reader must therefore translate the phrase into Aristophanic language." Vol. 2, p. 534.

3. Cited in Ernest Jones, *The Life and Work of Sigmund Freud*, Vol. I, New York, 1953, pp. 40–41.

4. Ludwig Büchner, *Force and Matter*, translated from the German, London, 1884, pp. 424–432.

5. Thomas H. Huxley, "On the Hypothesis that Animals Are Automata, and Its History," in G. N. A. Vesey, *Body and Mind*, London, 1964.

6. Thomas H. Huxley, *Man's Place in Nature*, London, 1863.

7. Thomas H. Huxley, *Method and Results*, London, 1868, pp. 31–32.

8. G. J. Allman, "Protoplasm and the Commonality of Life," in George Basalla ed., *Victorian Science*, New York, 1970, p. 261.

9. James Hinton, *Life in Nature*, London, 1862, p. xv.

10. Claude Bernard, *Introduction to the Study of Experimental Medicine*, New York, 1957, p. 60.

11. *Ibid.*, p. 63. Marshall McLuhan, *Understanding Media*, New York, 1964, has commented: "Perhaps most spectacular of all was Claude Bernard, whose *Introduction to the Study of Experimental Medicine* ushered science into *le milieu intérieur* of the body exactly at the time when the poets did the same for the life of perception and feeling," p. 159. McLuhan mentioned explicitly Baudelaire and Rimbaud.

12. Count de St. Léon, *Love and Its Hidden History*, Boston, 1869, pp. 80–90.

13. William Hooker, *Physician and Patient*, London, 1850, pp. 179–180.

14. Henry Maudsley, *The Physiology and Pathology of the Mind*, 1868; *Body and Mind*, 1870; *Life in Mind and Conduct*, 1902.

15. Maudsley, *Body and Mind*, pp. 34, 94, 95. Following Maudsley there appeared a number of studies of the mind-body relation particularly as it influenced physical and mental illness. Daniel Hack Tuke's enormous two-volume study appeared in London in 1872, *Illustrations of the Influence of the Mind upon the Body*. Another English physician, William Carpenter, devoted a study to the subject in 1874 with a textbook that carried a revealing, though somewhat enigmatic, title—*Mental Physiology*.

16. This article was expanded into a full-length study in 1876, *Vorschule der Aesthetik*, which developed this method of explaining the aesthetic response.

17. Grant Allen, *Physiological Aesthetics*, New York, 1877, pp. 2, 29. For another aesthetics in this genre see Georges Hirth, *Physiologie der Kunst*, 1885. Hirth also attempted to determine whether the aesthetic response was caused by an innate "psychophysical organization" of the human sensory system.

18. George Santayana, *The Sense of Beauty*, New York, 1955, pp. 53, 56. Santayana wrote: ". . . for man all nature is a secondary object of sexual passion, and . . . to this fact the beauty of nature is largely due," p. 62. Nietzsche first wrote of "sublimated sexuality" in *Human, All-Too Human* in 1879 and then returned to the subject in *Beyond Good and Evil* (1886). Havelock Ellis speculated in 1898 that art was rooted in the erotic impulse. For other theories about the connection between the sexual impulse and art see Colin A. Scott, "Sex and Art," *The American Journal of Psychology*, VII, January 1, 1896; Gustav Naumann, *Geschlecht und Kunst*, Leipzig, 1899; and Wilhelm Bölsche, *Das Liebesleben in der Natur*, Leipzig, 1911. Henri Ellenberger discussed other writers who derive man's sense of beauty from the sexual impulse—Steinthal, Möbius, Hirn, and Rémy de Goncourt—in *The Discovery of the Unconscious*, New York, 1970, p. 302.

19. F. N. Fowler, *How to Learn Phrenology*, London, 1881, p. 30.

20. Alfred T. Story, "A Chapter on Noses," *Phrenological Magazine*, London, 1881.

21. Arthur Cheetham, *Character Reading Practically Explained*, Manchester, 1893, p. 4.

22. Genesis III: 16.

23. Quoted in Hayword Haggard, *Devils, Drugs and Doctors*, New York, 1929, p. 116.

24. Abbé Bougard, *Le Christianisme dans les temps present*, 1877, cited by Jules Rochard, "Le Douleur," *Revue des deux mondes*, April, 15, 1889, p. 830.

25. *Ibid.*, p. 844.

26. Lewis Mumford, *Technics and Civilization*, New York, 1934, pp. 34–35.

CHAPTER 8

1. Robert Buchanan, *The Fleshly School of Poetry*, London, 1873, pp. 3, 15, 21, 44, 87.

2. Emile Zola, "The Experimental Novel," in Eugen Weber ed., *Paths to the Present*, New York, 1965, p. 171.

3. Quoted in Frederick Hemmings, *Emile Zola*, Oxford, 1953, p. 28.

4. *Ibid.*, p. 141.

5. Emile Zola, *Les Rougon-Macquart*, Vol. III, Paris, 1964, p. 1440.

6. *Ibid.*, Vol. II, Paris, 1961, pp. 1118–1119.

7. *Ibid.*, pp. 1269–1270.

8. *Ibid.*, p. 1485.

9. For studies that explicitly criticize the sexuality in Zola's work see Dr. Pascal, *Das sexuelle Problem in der modernen Literatur*, Berlin, 1890, p. 12; Max Nordau, *Degeneration*, translated from the German, New York, 1895; Leo Berg, *Das sexuelle Problem in Kunst und Leben*, Berlin, 1891, p. 10; Adolf Bartels, *Geschlechtsleben und Dichtung*, Berlin, 1906, p. 8.

10. Zola, *Rougon-Macquart*, II, p. 1314.

11. *Ibid.*, III, p. 1453.

12. Friedrich Nietzsche, *The Birth of Tragedy and the Genealogy of Morals*, translated from the German, New York, 1956, p. 299.

13. *Ibid.*, p. 128.

14. Friedrich Nietzsche, *Beyond Good and Evil*, translated by Walter Kaufmann, New York, 1966, Aphorism #2.

15. *Ibid.*, Aphorism #10.

16. *Ibid.*, Aphorism #33.

17. *Ibid.,* Aphorism #188.

18. *Ibid.,* Aphorism #225.

19. Nietzsche, *Genealogy,* p. 298.

20. Friedrich Nietzsche, *The Portable Nietzsche,* translated by Walter Kaufmann, New York, 1954, p. 203.

21. Nietzsche, *Genealogy,* p. 166.

22. *Ibid.,* p. 167.

23. *Ibid.,* p. 261.

24. Nietzsche, *Portable Nietzsche,* p. 188. Georg Lukacs has offered a similar evaluation of Nietzsche's philosophy of the body. "Ever since Nietzsche, the body *(Leib)* has played a leading role in bourgeois philosophy. The new philosophy [a synthesis of idealism and materialism] needs formulas which recognize the primary reality of the body and the joys and dangers of bodily existence, without, however, making any concessions to materialism," from George Novack ed., *Existentialism Versus Marxism,* New York, 1966, p. 136.

CHAPTER 9

1. Cited by Jill Conway, "Stereotypes of Femininity in a Theory of Sexual Evolution," in Martha Vicinus ed., *Suffer and Be Still,* Bloomington, Indiana, 1972, p. 141.

2. Patrick Geddes, *The Evolution of Sex,* London, 1889.

3. John S. Haller and Robin M. Haller surveyed a number of these theories in *The Physician and Sexuality in Victorian America,* Chicago, 1974, pp. 48–61. I am indebted to the Hallers for a number of references on theories about gender and family relations which I discuss in this (and the following) chapter.

4. Paul Möbius, *Ueber den physiologischen Schwachsinn des Weibes,* Leipzig, 1901.

5. Jules Michelet, *L'Amour,* Paris, 1858, p. 50.

6. Leo Berg, *Das sexuelle Problem in Kunst und Leben,* Berlin, 1891, p. 63.

7. John A. Smith, *The Mutations of Earth,* 1846, cited by Haller, *Physician and Sexuality,* p. 4.

8. Eduard Fuchs, *Illustrierte Sittengeschichte vom Mittelalter bis zur Gegenwart,*

Erganzungsband, Munich, 1912, p. 68. Fuchs refers to an original article in *Geschlecht und Gesellschaft*, Bd. II, Heft. 3.

9. Quoted by Elaine and English Showalter, "Victorian Women and Menstruation," *Victorian Studies*, September 1970, p. 85.

10. Michelet, *L'Amour*, p. 52.

11. Showalter, *Victorian Women*, p. 87.

12. Rosa Mayreder, *Zur Kritik der Weiblichkeit*, Jena, 1905, pp. 17–18.

13. Jos van Ussel, *Sexualunterdrückung: Geschichte der Sexualfeindschaft*, Hamburg, 1970, argued that "while in previous centuries the greater potency of women was widely acknowledged [in the nineteenth century] women's orgasms are denied or viewed as perverse. The consequence was that many women believed themselves to be frigid," p. 196. Milton Rugoff, *Prudery and Passion*, New York, 1971, cites as evidence of the denial of female sexuality in mid-nineteenth-century America the view of a New York physician that "without stimulation of drink, destitution, or seduction the full force of sexual desire is seldom known to virtuous women," p. 46.

14. William Acton, *The Function and Disorders of the Reproductive Organs*, London, 1857.

15. Alexander Walker, *Intermarriage*, 1850, p. 256, and Rufus Griswold, *Some Observations on the Physiology of Coitus from the Female Side of the Matter*, p. 447. Both sources cited by Haller, *Physician and Sexuality*, pp. 100–101.

16. George Napheys, *The Transmission of Life*, Philadelphia, 1871, pp. 173–174.

17. From *Rees' Cyclopedia*, undated. Cited by Peter T. Cominos, "Late-Victorian Sexual Respectability and the Social System," *International Review of Social History*, Vol. VIII, 1963, p. 23, who got the reference from Havelock Ellis, *Studies in the Psychology of Sex*.

18. Alfred Hegar, *Der Geschlechtstrieb*, Stuttgart, 1894, p. 20.

19. Van Ussel, *Sexualunterdrückung*, p. 199, mentions specifically the "physiologist" Magendie and the *"Frauenärzte"* Busch, Theopold (1873) and Otto Adler. François Magendie (1783–1855) was influential in the early nineteenth century, and perhaps Ussel had him in mind from a later edition of his work in the 1870s. I was unable to ascertain the identity of either Busch or Theopold.

20. A Physician, *Satan in Society*, Cincinnati, 1881, p. 155.

21. Richard Ungewitter, *Nacktheit und Kultur*, 1919.

22. Havelock Ellis, "Auto-Erotism," *The Alienist and Neurologist*, April, 1898, pp. 9–11.

23. A Physician, *Satan in Society*, p. 106.

24. Pierre Garnier, *Onanisme seul et à deux sous toutes ses formes et leur conséquences*, Paris, 1894, p. 350.

25. A. Dechambre ed., *Dictionnaire encyclopédique des sciences médicales*, Paris, 1881, p. 384.

26. Garnier, *Onanisme*, p. 359.

27. Paul Flechsig, "Zur gynaekologischen Behandlung der Hysterie," *Neurologisches Centralblatt*, Vol. 19–20, 1884.

28. Alfred Hegar, *Der Zusammenhang der Geschlechtskrankheiten mit nervosen Leiden und die Castration bei Nervosen*, Stuttgart, 1885.

29. John Boyle O'Reilly, *Athletics and Manly Sport*, Boston, 1890, p. 102.

30. Dudley Allen Sargent, *An Autobiography*, Philadelphia, 1927, p. 202.

31. "Remarks and Comments," *Mind and Body*, June, 1894, p. 16.

32. Henry Thétard, *Coulisses et secrets du cirque*, Paris, 1934, p. 21.

33. Henry Thétard, *Le Merveilleuse histoire du cirque*, Vol. II, Paris, 1947: "The success of the Rassos coincided with the great renaissance of sports and physical culture. The fin-de-siècle public went wild over a number of Hercules. . . . The century culminated in a debauch of spectacular athletics," p. 72.

34. Bernarr Macfadden, *The Virile Powers of Superb Manhood*, New York, 1900, p. 6.

35. Fuchs, *Illustrierte Sittengeschichte*, Vol. III, pp. 133–135.

36. Claude-François Lallemand, *Des pertes seminales involontaires*, 1836, cited by Haller, *Physician and Sexuality*, p. 211.

37. J. L. Milton, *On the Pathology and Treatment of Spermatorrhoea*, New York, 1887, pp. 397–401.

38. Haller, *Physician and Sexuality*, p. 215, cites as a source: Roberts Bartholow, *On Spermatorrhoea: Its Causes, Symptomatology, Pathology, Prognosis, Diagnosis, and Treatment*, New York, 1886, pp. 85–86.

39. Pierre Garnier, *Impuissance physique et morale*, Paris, 1883, p. 17.

40. *Ibid.*, pp. 485-487.

41. A Physician, *Satan in Society*, p. 149.

42. Victor Vecki, *The Pathology and Treatment of Sexual Impotence*, translated from the German, London, 1901, p. 88.

43. Haller, *Physician and Sexuality*, p. 23.

44. *Ibid.*, p. 15.

45. Emil Peters, *Die Wiedergeburt der Kraft*, 1908, p. 247.

46. Sylvanus Stall, *What a Young Man Ought to Know*, London, 1897, pp. 104-105.

CHAPTER 10

1. John S. Haller and Robin M. Haller, *The Physician and Sexuality in Victorian America*, Chicago, 1974, p. 115, mentions several such arguments in popular medical literature before 1870.

2. Robert Michels, *Die Grenzfragen der Geschlechtsmoral*, 1911, p. 122.

3. Ronald Pearsall, *The Worm in the Bud*, London, 1969, p. 224.

4. George Napheys, *The Transmission of Life*, 1878, p. 107.

5. Carl Gelsen, *Die Hygiene der Flitterwochen*, Berlin, 1889, pp. 28-30.

6. A. Schotta "Über Erziehung der Brüste zum Stillen," in Karl Vanselow ed., *Sexualreform*, 1907, p. 56.

7. Alexander von Gleichen-Russwurm, *Geselligkeit, Sitten und Gebräuche der europäischen Welt: 1789-1900*, Stuttgart, 1910, p. 110, cited by William Treue, *Kleine Kulturgeschichte des deutschen Alltags*, Potsdam, 1942, pp. 310-311.

8. Georgiana B. Kirby, *Transmission; or, Variation of Character Through the Mother*, New York, 1877, p. 11.

9. Haller, *Physician and Sexuality*, discusses several American writers who speculated about the various nefarious consequences of a pregnant woman's engaging in sexual relations (Wood-Allen, Kellogg, Fowler, etc.) pp. 131-132.

10. A. Coriveaud, *Le Lendemain du mariage*, Paris, 1884, p. 117.

11. A Physician, *Satan in Society*, Cincinnati, 1881, p. 209.

12. Eduard von Hartmann, *The Sexes Compared*, translated from the German, London, 1895, p. 12.

13. Otto Weininger, *Sex and Character*, translated from the German, New York, 1906, p. 223.

14. Hermann Swoboda, *Studien zur Grundlegung der Psychologie*, Leipzig, 1905, p. 33.

15. Samuel Butler, *Unconscious Memory*, London, 1880, p. 19.

16. Gabriel Compayré, *The Intellectual and Moral Development of the Child*, translated from the French, New York, 1896, p. 258.

17. Henrik Ibsen, *Ghosts and Three Other Plays*, New York, 1966, p. 163.

18. Cited in Steven Marcus, *The Other Victorians*, New York, 1964, p. 27.

19. D. G. M. Schreber, *Anthropos. Der Wunderbau des menschlichen Organismus*, Leipzig, 1859, p. 95.

20. D. P. Schreber, *Memoirs of My Nervous Illness*, translated from the German edition of 1903, London, 1955.

21. See William G. Niederland, "The 'Miracled-Up' World of Schreber's Childhood," *Psychoanalytical Study of the Child*, Vol. 14, 1949; William G. Niederland, "Schreber: Father and Son," *The Psychoanalytic Quarterly*, Vol. 28, 1959; and William G. Niederland, "Further Data and Memorabilia Pertaining to the Schreber Case," *International Journal of Psychoanalysis*, Vol. 44, 1963.

22. D. P. Schreber, *Memoirs*, p. 131.

23. Niederland, "Further Data," p. 201.

24. D. G. M. Schreber, *Kallipädie oder Erziehung zur Schönheit durch naturgetreue und gleichmässige Förderung normaler Körperbildung* (Education Toward Beauty by Natural and Balanced Furtherance of Normal Body Growth), Leipzig, 1858, pp. 137–138, quoted in Morton Schatzman, "Paranoia or Persecution: The Case of Schreber," *History of Childhood Quarterly*, Summer 1973, p. 78.

25. The metaphor of the body as "spokesman" was suggested in an article by Erik H. Erikson, "The First Psychoanalyst," in Benjamin Nelson ed., *Freud and the 20th Century*, New York, 1957. "It fits our image of those Victorian days—a time when children in all and women in most circumstances were to be seen but not heard—that the majority of symptoms would prove to lead back to events when violently aroused

affects (love, sex, rage, fear) had come into conflict with narrow standards of propriety and breeding. The symptoms, then, were delayed involuntary communications: using the whole body as spokesman, they were saying what common language permits common people to say directly: 'He makes me sick,' 'She pierced me with her eyes' . . ." p. 85.

26. René Spitz, "Authority and Masturbation," *The Psychoanalytic Quarterly*, Vol. 21, 1952, p. 499.

27. For a demonstration of the growing awareness of child sexuality in the late nineteenth century see my article, "Freud and the Discovery of Child Sexuality," *History of Childhood Quarterly*, Summer 1973, pp. 117–141.

28. G. Stanley Hall, *Adolescence*, New York, 1904, Vol. II, p. 438.

29. Haller, *Physician and Sexuality*, pp. 207–208.

30. O. S. Fowler, *Love and Parentage Applied to the Improvement of Offspring: Including Important Directions and Suggestions to Lovers and the Married Concerning the Strongest Ties and the Most Sacred and Momentous Relations of Life*, New York, 1846, p. 74.

CHAPTER 11

1. Erik Nordenskiöld, *The History of Biology*, New York, 1928, p. 543.

2. Wilhelm Bölsche, *Love-Life in Nature: The Story of the Evolution of Love*, translated from the German, New York, 1926, Vol. I, pp. 18, 28.

3. Wilhelm Bölsche, *Das Liebesleben in der Natur*, Leipzig, 1911, p. 481.

4. Bölsche, *Love-Life*, pp. 43–51.

5. Quoted in Havelock Ellis, *Studies in the Psychology of Sex*, Philadelphia, Vol. VI, p. 135.

6. Max Dessoir, "Zur Psychologie der Vita Sexualis," *Allgemeine Zeitschrift für Psychiatrie*, Vol. 50, 1894, p. 942.

7. Albert Moll, *Untersuchungen über die Libido Sexualis*, Berlin, 1898, p. 23.

8. Hans Rau, *Der Geschlechtstrieb und seine Verirrungen*, Berlin, 1903, p. 13.

9. Friedrich Scholz, *Die moralische Anästhesie*, Leipzig, 1904, p. 105.

10. Ernest Chambard, *Du Somnambulisme en général*, Paris, 1881.

11. P. Kerval, "Sociétés savantes," *Archives de Neurologie*, Vol. 6, 1883, p. 131.

12. Albert Pitres, *Des Zones hystérogènes et hypnogènes*, Paris, 1885.

13. Moll, *Untersuchungen*, p. 93.

14. Iwan Bloch, *Beiträge zur Aetiologie der Psychopathia Sexualis*, Dresden, 1902, Vol. II, pp. 43-44.

15. Otto Stoll, *Das Geschlechtsleben in der Volkerpsychologie*, 1908.

16. Edward M. Brecher, *The Sex Researchers*, Boston, 1969, p. 18.

17. Françoise Delisle, *Friendship's Odyssey*, London, 1946.

18. Hermann Rohleder, *Die Masturbation*, Berlin, 1899.

19. Alexander Chamberlain, *The Child: A Study in the Evolution of Man*, New York, 1911, p. 402.

20. Otto Weininger, *Sex and Character*, translated from the German, New York, 1906, pp. 100-101.

21. *Ibid.*, p. 266.

22. *Ibid.*, pp. 233, 250, 347.

23. *Ibid.*, p. 311.

24. August Forel, *The Sexual Question*, translated from the German, New York, 1908, pp. 503-508.

25. *Ibid.*, p. 512.

CHAPTER 12

1. Richard von Krafft-Ebing, *Psychopathia Sexualis*, translated from the German, New York, 1965, pp. xiii, 1.

2. *Ibid.*, pp. 8-9.

3. *Ibid.*, p. 145.

4. Jean Charcot and Valentin Magnan, "Inversion du sens genital," *Archive de Neurologie*, 1882.

5. Alfred Binet, "Le Fétichisme dans l'amour," *Revue Philosophique*, 1887, introduced the word "fetishism" to the French public to describe those sexual aberrations in which the usual goal of heterosexual intercourse is abandoned for substitute objects such as other parts of the body or articles of clothing.

6. See chapter 5 above on fetishes involving smell.

7. Wilhelm Fliess, *Die Beziehungen zwischen Nase und weibliche Geschlechtsorganen*, Leipzig, 1897.

8. Krafft-Ebing, *Psychopathia Sexualis*, p. 37.

9. *Ibid.*, p. 65.

10. Leopold von Sacher-Masoch, *Venus in Furs*, translated from the German, London, 1971, p. 228.

11. Conrad Haemmerling, *Kultur und Sittengeschichte der neuesten Zeit*, Dresden, 1929, Vol. III, pp. 340, 346.

12. See Eugen Dühren (pseud. for Iwan Bloch), *Sexual Life in England*, translated from the German, London, 1938. The German original is even more complete. Ronald Pearsall, *The Worm in the Bud*, London, 1969, discusses the subject pp. 404–421.

13. H. Montgomery Hyde, *The Love that Dared Not Speak Its Name*, Boston, 1970, p. 144.

14. *The Yellow Book* was an *avant garde* periodical that first appeared in 1894. But even though Wilde was excluded from publishing in it, when he was arrested in 1895 he carried a yellow-covered French novel under his arm as he went into custody, and the newspapers played up the mistaken fact that when he was arrested he had a copy of *The Yellow Book*. The publisher was then pressured to exclude the work of Aubrey Beardsley from the journal, because he had done erotic drawings for Wilde's shocking play, *Salomé*. Sales fell and the journal expired in 1897. Havelock Ellis's *Sexual Inversion* was delayed for a time because Horatio Brown, the literary executor of its co-author, John Addington Symonds, panicked following the Wilde trial and insisted that Symonds's name be removed from the title page and all of his contributions withheld. The version that appeared under Ellis's name alone was very much different from the German edition that had appeared in 1896. Edward Carpenter's *Homogenic Love and Its Place in a Free Society* was greeted by an icy silence following its publication in 1895. Carpenter helped counter the fear created by the Wilde trial with the publication of *Love's-Coming-Of-Age* in 1896.

15. Havelock Ellis explained in *Sexual Inversion*: "The Oscar Wilde trial, with its wide publicity . . . appears to have generally contributed to give definiteness and self-consciousness to the manifestations of homosexuals, and to have aroused inverts to take up a definite attitude," p. 212.

16. See H. Montgomery Hyde, *Oscar Wilde*, Middlesex, 1962, p. 60, for an account of Wilde's homosexual activities.

17. Paul Moreau, *Des Aberrations du sens génésique*, Paris, 1880.

18. Benjamin Tarnowsky, *L'Instinct sexuel et ses manifestations morbides*,

Paris, 1904; Krafft-Ebing, *Psychopathia Sexualis*, 1886; and Albert Moll, *Die konträre Sexualempfindung*, 1891.

19. Karl Ulrichs, *Memnon: Die Geschlechtsnatur des mannliebenden Urnings*, Schleiz, 1868.

20. Phyllis Grosskurth, *The Woeful Victorian: A Biography of John Addington Symonds*, New York, 1964, p. 86.

21. Donald Cory ed., *Homosexuality: A Cross Cultural Approach*, New York, 1956, p. 98. The Napoleonic Code abolished all punishment for homosexuality provided that it took place in private between consenting adults.

22. Ellis, *Sexual Inversion*, p. 184. Henri Ellenberger, *The Discovery of the Unconscious*, New York, 1970, p. 204, has traced the theory of bisexuality to early German Romantic philosophy. It became the focus of a heated argument around the turn of the century when Wilhelm Fliess accused Otto Weininger of having plagiarized the theory of bisexuality from him through the unwitting mediation of Freud and his student, Hermann Swoboda. For Fliess's frantic accusations see Wilhelm Fliess, *Im eigener Sache: gegen Otto Weininger und Hermann Swoboda*, Berlin, 1906.

23. Edward Carpenter, "The Intermediate Sex" in Cory, *Homosexuality*, pp. 202, 203.

24. G. W. Ireland, *André Gide*, Oxford, 1970, p. 312.

25. André Gide, *Corydon*, translated from the French, New York, 1950, p. 94.

26. *Ibid.*, p. 107.

27. H. Stuart Hughes, *Consciousness and Society*, New York, 1961, p. 363.

CHAPTER 13

1. See Steven Marcus, *The Other Victorians*, New York, 1964; Charles Rosenberg, "Sexuality, Class and Role in 19th-Century America," *American Quarterly*, May 1973; and John S. Haller and Robin M. Haller, *The Physician and Sexuality in Victorian America*, Chicago, 1974.

2. G. Stanley Hall, *Adolescence*, New York, 1916, Vol. I, p. 469.

3. R. Ussher, *Neo-Malthusianism*, 1897, cited by Ronald Pearsall, *The Worm in the Bud*, London, 1969, p. 278.

4. Francis Place, *Illustrations and Proofs of the Theory of Population*, 1822.

5. Pearsall, *Worm in the Bud*, p. 274.

6. John Humphrey Noyes, *Male Continence*, New York, 1872, p. 8. For more on Noyes see Maren Lockwook Carden, *Oneida; Utopian Community to Modern Corporation*, New York, 1971.

7. Alexander Peyer, *Der unvollständige Beischlaff (Congress Interruptus, und seine Folgen beim mannlichen Geschlechte)*, Stuttgart, 1890.

8. Iwan Bloch, *Das Sexualleben unserer Zeit*, Berlin, 1907, pp. 758–760.

9. George Napheys, *The Transmission of Life: Counsels on the Nature and Hygiene of the Masculine Function*, Philadelphia, 1878, p. 149.

10. *Ibid.*, pp. 173–174.

11. Mary Wood-Allen and Sylvanus Stall, *What a Young Girl Ought to Know*, Philadelphia, 1897, pp. 105–106.

12. *Ibid.*, pp. 108–109.

13. *Ibid.*, p. 110.

14. *Ibid.*, pp. 111, 115.

15. *Ibid.*, pp. 133–134.

16. Frank Wedekind, "Frühlings Erwachen," in *Prosa Dramen Verse*, Munich, n.d., pp. 270–271.

17. For an excellent summary of these ideas see John S. Haller, Jr., "From Maidenhead to Menopause: Sex Education for Women in Victorian America," *Journal of Popular Culture*, Summer, 1972.

18. J. H. Kellogg, *Ladies' Guide to Health and Disease*, Iowa, 1883, quoted by Haller, *Ibid.*, p. 52.

19. Haller, *Ibid.*, p. 55.

20. Seved Rebbing, *Die sexuelle Hygiene und ihre ethischen Konsequenzen*, translated from the Swedish, Leipzig, 1890.

21. Pierre Garnier, *Le Mal d'amour, contagion, préservatifs et remèdes*, Paris, 1891, and *Le Mariage dans ses devoirs, ses rapports et ses effets conjugaux*, Paris, 1908.

22. Anna Fischer-Dückelmann, *Die Frau als Hausarztin*, Stuttgart, 1901, p. 127.

23. *Ibid.*, p. 230.

24. *Ibid.*, p. 234.

25. Bloch, *Sexualleben*, p. 778.

26. Theodore van de Velde, *Ideal Marriage*, New York, 1926, pp. 7, 8, 26, 179, 164. *Ideal Marriage* went through forty-two printings in Germany between 1926 and 1932.

CHAPTER 14

1. Sigmund Freud, "On the Universal Tendency to Debasement in the Sphere of Love," *The Standard Edition of the Complete Psychological Works of Sigmund Freud*, translated from the German under the General Editorship of James Strachey, Vol. 11, p. 189. I have not been able to ascertain the precise saying by Napoleon that Freud had in mind. One possibility is a quotation attributed to Napoleon in 1815 during a conversation aboard the Bellerophon: "Destiny must be fulfilled—that is my chief doctrine." Quoted in J. Christopher Herold, *The Mind of Napoleon*, New York, 1955, p. 40.

2. Freud, "The Dissolution of the Oedipus Complex," *Standard Edition*, Vol. 19, p. 178.

3. Freud, "Some Points For a Comparative Study of Organic and Hysterical Motor Paralysis," *Standard Edition*, Vol. 1, pp. 169–170.

4. Freud, "The Neuro-Psychoses of Defense," *Standard Edition*, Vol. 3, p. 49.

5. *Ibid.*, p. 60.

6. Freud, "Five Lectures on Psycho-Analysis," *Standard Edition*, Vol. 11.

7. Freud, *Beyond the Pleasure Principle*, quoted by Felix Deutsch, *On the Mysterious Leap From the Mind to the Body*, New York, 1959, p. 31.

8. Freud, "Studies in Hysteria," *Standard Edition*, Vol. 2, p. 86.

9. *Ibid.*, p. 152.

10. Freud, "Fragment of an Analysis of a Case of Hysteria," *Standard Edition*, Vol. 7, p. 52.

11. Freud, "The Psycho-Analytic View of Psychogenic Disturbance of Vision," *Standard Edition*, Vol. 11.

12. Freud, "Introductory Lectures on Psycho-Analysis," *Standard Edition*, Vol. 15, pp. 176–177.

13. Sigmund Freud, *A General Introduction to Psychoanalysis*, New York, 1962, p. 269. Deutsch refers to this passage in *Mysterious Leap*, p. vii, and uses the metaphor in the title of his study of Freud.

14. Freud, "The Interpretation of Dreams," *Standard Edition*, Vol. 5, p. 351.

15. Ernst Kris ed., *The Origins of Psycho-Analysis*, New York, 1954, p. 188.

16. Although Freud is particularly well known for his catalog of sexual symbols, thinking on the subject goes back to a seminal work by the German philosopher Karl Albert Scherner in 1861 which focused heavily on the corporeal origins of dream symbols. Scherner surveyed the somatic sources of dreams and speculated how these stimuli are transformed into symbols: a house symbolizes the body, water symbolizes the urinary process, and fire symbolizes respiration. In a section on dreams stimulated by intestinal activity, he reconstructed a typical dream with the appropriate symbolic representations. The dream begins with a meal, after which the dreamer walks out into the night (darkness represents the stomach). After a pause in the square (the stomach) he begins to make his way out on a small road. This symbolic representation of the lower intestines becomes progressively narrower, winding, and muddy. At the end of the road the mud becomes so thick as to make the road impassable, and the traveler encounters a tangle of branches which Scherner interpreted as representing the anal sphincter. Karl Albert Scherner, *Das Leben des Traumes*, Berlin, 1861, p. 175.

17. Freud, "Introductory Lectures on Psycho-Analysis," *Standard Edition*, Vol. 15, p. 153.

18. Freud, *Interpretation of Dreams*, p. 354.

19. *Ibid.*, p. 355.

20. *Ibid.*, p. 399.

21. Freud, "The Claims of Psycho-Analysis to Scientific Interest," *Standard Edition*, Vol. 13, p. 180.

22. Freud, "Character and Anal Erotism," *Standard Edition*, Vol. 9, pp. 169–170.

23. *Ibid.*, p. 175.

24. Freud, *Three Essays on the Theory of Sexuality*, New York, 1962, p. 88.

25. Freud, "Introductory Lectures on Psycho-Analysis," *Standard Edition*, Vol. 16, p. 318.

26. Freud, "Some Psychical Consequences of the Anatomical Distinction Between the Sexes," *Standard Edition*, Vol. 19, pp. 256–257.

27. *Ibid.*, pp. 257–258.

28. Freud, "Female Sexuality," *Standard Edition*, Vol. 21, p. 228.

29. Freud, "Three Essays on the Theory of Sexuality," *Standard Edition*, Vol. 7, p. 243.

30. Freud, "'Civilized' Sexual Morality and Modern Nervous Illness," *Standard Edition*, Vol. 9, p. 191.

31. Freud, "On the Universal Tendency to Debasement in the Sphere of Love," *Standard Edition*, Vol. 11, pp. 184-185.

CHAPTER 15

1. George L. Mosse, *The Culture of Western Europe*, New York, 1961, p. 277.

2. H. Stuart Hughes, *Consciousness and Society: The Reorientation of European Social Thought, 1890-1930*, New York, 1968, analyzed this historical change. The generation of social scientists before the war sought to "comprehend the newly recognized disparity between external reality and the internal appreciation of that reality," p. 16. Hughes considered "the revolt against positivism" explicitly in Chapter 2, where he examined the study of "irrational motivation in human conduct," p. 35.

3. Emanuele Gallo, *La guerra e la sua ragione sensuale*, Torino, 1912. I have been unable to locate a copy of this book.

4. Magnus Hirschfeld, *Sittengeschichte des ersten Weltkrieges*, 1929. I used the Second Edition, Verlag Karl Schustek, Hanau, n.d., p. 51. A different book that covers similar material appeared during the Second World War, Magnus Hirschfeld, *The Sexual History of the World War*, New York, 1941. I have relied on the original German edition in this chapter.

5. From an editorial of 1903 in a Berlin periodical, cited by Bruno Grabinski, *Weltkrieg und Sittlichkeit*, Hildesheim, 1917, p. 21.

6. Erhardt W. Eberhard, *Die Frauenbewegung und ihre erotischen Grundlagen*, Vienna, 1924. On the image of the *femme fatale* see also Mario Praz, *The Romantic Agony*, Cleveland, 1956, Chapter IV, "La Belle Dame Sans Merci," and Philippe Jullian, *Dreamers of Decadence*, translated from the French, New York, 1971, *passim*.

7. Helene Stöcker, *Geschlechtspsychologie und Krieg*, Berlin, 1916, p. 3, cites the article by Karl Scheffler that appeared in the *Vossische Zeitung*, No. 463, September 10, 1915. Gustav Vorberg introduced the phrase "Stahlbad der Nerven" in *Das Geschlechtsleben im Weltkriege*, Munich, 1918, p. 10.

8. Henry Miller, *The Cosmological Eye*, New York, 1939, pp. 119-120.

9. George Orwell, *Homage to Catalonia*, New York, 1952, p. 44.

10. Grabinski, *Weltkrieg*, p. 90; P. Lissmann, *Die Wirkungen des Krieges auf das mannliche Geschlechtsleben*, Munich, 1919, p. 1.

11. H. C. Fischer and E. X. Dubois, *Sexual Life During the World War*, London, 1937, pp. 249–250.

12. Mary Agnes Hamilton, *Our Freedom*, p. 251, quoted by Arthur Marwick, *The Deluge: British Society and the First World War*, New York, 1965, p. 108.

13. Lissmann, *Wirkungen des Krieges*, p. 6.

14. Hirschfeld, *Sittengeschichte*, p. 61.

15. Fischer, *Sexual Life*, p. 285.

16. Magnus Hirschfeld, "War Brothels," in Victor Robinson ed., *Morals in Wartime*, New York, 1943, p. 129.

17. *Ibid.*, p. 125.

18. Cited by Hirschfeld, *Ibid.*, p. 131.

19. Marwick, *The Deluge*, p. 107.

20. Fischer, *Sexual Life*, p. 139, cites a work by a German physician, Müller, *Effect of the War on the Feminine Organism*. I have not been able to identify the complete reference.

21. For an account of these and other atrocities see Grabinski, *Weltkrieg*, Chapter 5.

22. Erich Wulffen, *Woman as a Sexual Criminal*, translated from the German, New York, 1934, ". . . during the World War women who stripped and plundered dead bodies on the field of battle would cut off the penises of the dead or still living men," p. 186.

23. Grete Meisel-Hess, *Krieg und Ehe*, 1915, cited by Hirschfeld, *Sittengeschichte*, p. 98.

24. Hirschfeld, *Ibid.*

25. Emil Gustav Paulk, *Die Manneslehre*, 1918, p. 113.

26. *Ibid.*, Chapter 8.

27. E. Heinrich Kisch, *Die sexuelle Untreue der Frau*, 1918, pp. 88–89.

28. Lewis Mumford, *Technics and Civilization*, New York, 1934, pp. 97–98.

CHAPTER 16

1. Quoted by Milton Rugoff, *Prudery and Passion*, New York, 1971, p. 123.
2. Karl Vanselow ed., *Sexualreform*, 1907, p. 279.
3. D. H. Lawrence, *Lady Chatterley's Lover*, New York, 1968, p. 4.
4. *Ibid.*, p. 55.
5. Kate Millett, *Sexual Politics*, New York, 1970, pp. 237–245. Millett interpreted the work of Lawrence, as well as that of Freud and Henry Miller, as a source of the counter-revolutionary sexual politics of the twentieth century which sought to subject women to sexual subordination by men. She criticized Freud's theory that women are anatomically determined to assume passive sex roles and forever yearn for the man's penis. Although she conceded that much of her criticism should be directed at "vulgar Freudianism," she argued that anti-feminism was inherent in Freud's own work and that Freud was "the strongest counter-revolutionary force in the ideology of sexual politics" of the twentieth century. In Lawrence, Freud's notions became "superb instruments for the perfect subjection of women," and Miller, she insisted, "unleashed and made fashionable" contempt for female sexuality. My evaluation of the contributions of these men are very much opposed to hers. I view them as pioneers in the study of human sexuality who contributed as much as anyone else in their time to the understanding and enjoyment of the human body. Their work is in some ways dated, but only because they addressed themselves to their current problems. They indeed shared some of the prejudices of their age and in some ways expressed views that now appear awkward and even a bit pernicious. But progress is never unambiguous, and to insist that it be uncompromised is to place requirements upon it that are impossible to satisfy. Freud, Lawrence, and Miller succeeded far better than any of their contemporaries in challenging the conspiracy of silence and in opening public forums for the discussion of human sexuality and the manifold problems it can engender. There is little value in assailing them for not supplying the answers that we want to hear or for not creating the kind of world that we would like to live in.
6. Lawrence, *Lady Chatterley's Lover*, p. 187.
7. *Ibid.*, pp. 134, 184.
8. *Ibid.*, p. 226.
9. *Ibid., p. 188.*

10. *Ibid.*, p. 241.

11. *Ibid.*, p. 331.

12. *Ibid.*, p. 334.

13. *Ibid.*, p. 331.

14. *Ibid.*, p. 275.

15. *Ibid.*, p. 254.

16. D. H. Lawrence, *The Later D. H. Lawrence*, New York, 1959, p. 444.

17. Henry Miller, *Tropic of Cancer*, New York, 1961, p. 5.

18. *Ibid.*, p. 39.

19. *Ibid.*, p. 41.

20. *Ibid.*, pp. 84–88.

21. *Ibid.*, pp. 126–127.

22. *Ibid.*, p. 90.

23. *Ibid.*, p. 222.

24. *Ibid.*, pp. 222–232.

CHAPTER 17

1. George L. Mosse, *The Crisis of German Ideology: Intellectual Origins of the Third Reich*, New York, 1964, p. 3.

2. Richard D. Mandell, *Nazi Olympics*, New York, 1971, pp. 10–12.

3. *Ibid.*, p. 12.

4. Hildegard Brenner, *Die Kunstpolitik des Nationalsozialismus*, Hamburg, 1963, p. 27.

5. *Ibid.*, pp. 31–33.

6. Hans Peter Bleuel, *Strength Through Joy: Sex and Society in Nazi Germany*, translated from the German, London, 1973, p. 182.

7. Quoted in George L. Mosse ed., *Nazi Culture: Intellectual, Cultural and Social Life in the Third Reich*, New York, 1966, p. 115.

8. Paul Schultze-Naumberg, *Modische Schönheit*, Berlin, 1937, p. 95, quoted in Bleuel, *Strength Through Joy*, pp. 187–188.

9. Wolfgang Willrich, *Säuberung des Kunsttempels*, Berlin, 1937, pp. 143–144. In this same genre see Paul Schultze-Naumberg, *Kunst und Rasse*, Munich, 1928.

10. Cited by Mosse, *Crisis*, p. 98.

11. Vacher de Lapouge, *Les Selections sociales*, 1876, advocated ensuring French racial purity by a eugenics program.

12. Mosse, *Crisis*, pp. 88–107.

13. Christian H. Stratz, *Die Rassenschönheit des Weibes*, Stuttgart, 1901, p. 3. Other works by Stratz revealing strong racist sexual aesthetics are *Die Frauen auf Java: Eine gynäkologische Studie*, Stuttgart, 1897; *Die Schönheit des weiblichen Körpers*, Stuttgart, 1898; and *Die Körperformen in Kunst und Leben der Japaner*, Stuttgart, 1902. This latter work makes some positive evaluations of the attitudes of the Japanese about the body and their sex ethic—a reversal of his more frequent interpretation of the superiority of European culture.

14. Quoted in Wilfred Daim, *Der Mann, der Hitler die Ideen gab*, Munich, 1958, pp. 124–128.

15. Lanz von Liebenfels ed., "Nackt- und Rassekultur im Kampf gegen Mucker- und Tschandalenkultur," *Ostara: Bücherei für Blonde und Mannesrechtler*, Heft. 66; and "Die Beziehungen der Dunklen und Blonden zur Krankheit, *Ibid.*, Heft. 67.

16. Quoted in Otto Friedrich, *Before the Deluge*, New York, 1972, p. 225.

17. Alfred Rosenberg ed., *Dietrich Eckart: Ein Vermächtnis*, Munich, 1928, quoted in Mosse, *Nazi Culture*, p. 77.

18. Hermann Paull, *Deutsche Rassenhygiene: Ein gemeinverständliches Gespräch über Vererbungslehre, Eugenik, Familie, Sippe, Rasse und Volkstum*, Görlitz, 1934, quoted in Mosse, *Nazi Culture*, p. 35.

19. Julius Streicher, "Artfremdes Eiweiss," *Deutsche Volksgesundheit aus Blut und Boden*, Nuremberg, 3 Jg. Nr. 1, 1935, p. 1.

20. H. Baur, E. Fischer, F. Lenz, *Menschlichen Erblehre und Rassenhygiene*, 1936, cited by K. Saller, *Die Rassenlehre des Nationalsozialismus in Wissenschaft und Propaganda*, Darmstadt, 1961, p. 69.

21. Magnus Hirschfeld, *Sittengeschichte des ersten Weltkrieges*, Hanau, n.d., discusses the role of Jewish scientists in the study of syphilis and cites as his source Herbert Lewandowski, *Les Enfers, domaine allemand*, p. 171–172. I was unable to locate the study by Lewandowski.

22. Adolf Hitler, *Mein Kampf*, translated from the German, Boston, 1962, p. 328.

23. *Ibid.*, p. 57.

24. *Ibid.*, p. 247.

25. *Ibid.*, p. 325.

26. *Ibid.*, p. 412.

27. *Ibid.*, p. 404.

28. Franz Neumann, *Behemoth: The Structure and Practice of National Socialism 1933–1944*, New York, 1942, p. 112.

29. *Ibid.*, p. 111.

30. Bleuel, *Strength Through Joy*, p. 192.

31. Alexander Mitscherlich, *Medizin ohne Menschlichkeit*, Document No. 185, cited by Saller, *Rassenlehre*, p. 145.

32. Bleuel, *Strength Through Joy*, p. 137.

33. Wilhelm Reich, *The Function of the Orgasm*, New York, 1942, p. 4, quoted by Paul A. Robinson, *The Freudian Left*, New York, 1969, p. 13.

34. Reich, *Ibid.*, p. 59.

35. Wilhelm Reich, "Dialectical Materialism and Psychoanalysis," in Lee Baxandall ed., *Sex-Pol Essays 1929–1934*, p. 72.

36. Wilhelm Reich, "The Imposition of Sexual Morality," in Baxandall, ed., *1929–1934, Sex-Pol Essays*, pp. 89–249.

37. *Ibid.*, p. 245.

38. Reich, *Function*, pp. 159–167.

CHAPTER 18

1. Maurice Merleau-Ponty, "Man and Adversity," *Signs*, translated from the French, Northwestern University Press, 1964, pp. 226–227.

2. Henri Bergson, *Essai sur les données immédiates de la conscience*, Paris, 1888. I am indebted to Richard M. Zaner, *The Problem of Embodiment*, The Hague, 1964, for an interpretation of Bergson's contribution to the philosophy of embodiment. Zaner concluded: "Although neither Marcel, Sartre nor Merleau-Ponty has apparently noted this, it was Bergson who

first saw, with great insight, the genuine significance and peculiarity of the body, and the necessity of re-formulating the question of the relations between mind and body in terms of an analysis of the human body," p. 243.

3. Henri Bergson, *Matière et mémoire*, Paris, 1953, pp. 11, 16, cited in Zaner, *Ibid.*, pp. 245–246.

4. These terms are introduced in Part II of Edmund Husserl's *Ideen zu einer reinen Phänomenologie*, 1913.

5. Gabriel Marcel, *Metaphysical Journal*, translated from the French, Chicago, 1952, p. 126.

6. *Ibid.*, pp. 274, 315.

7. Michael Gelven, *A Commentary on Heidegger's "Being and Time,"* New York, 1970, wrote that Heidegger establishes the validity of "asking, not what kind of thing a being is, but what it means to be at all," p. 18.

8. Merleau-Ponty, *Signs*, pp. 226–230.

9. Jean-Paul Sartre, *Nausea*, translated from the French, New York, 1964, pp. 98–99, 129–131.

10. Jean-Paul Sartre, *Being and Nothingness*, translated from the French by Hazel E. Barnes, New York, 1956, p. 318.

11. *Ibid.*, p. 338.

12. *Ibid.*, p. 358.

13. This three-stage construction of disease awareness is made explicit in a footnote by the translator and editor, Hazel Barnes, p. 356.

14. Sartre, *Being and Nothingness*, p. 364.

15. *Ibid.*, p. 260.

16. *Ibid.*, pp. 269–271.

17. *Ibid.*, p. 377.

18. *Ibid.*, pp. 382–383.

19. *Ibid.*, pp. 386, 389.

20. *Ibid.*, p. 390.

21. *Ibid.*, p. 398.

22. *Ibid.*, p. 401.

23. *Ibid.*, pp. 403–404.

Index

Abortion, 165, 195, 201, 235
Abreaction, 172
Acton, William, 43, 99, 100, 117, 157
Adler, Alfred, 231
Adolescence (Hall), 154
Adultery, 4, 140, 201, 203–05, 232
Aesthetics, 74, 252
Aesthetics of the Ugly (Rosenkranz), 25–26
Against Nature (Huysmans), 53–55
Aggression, 188–89
Alienation, 57–58; of labor, 60–61; sexual, 56–62, 269
All Quiet on the Western Front (Remarque), 195
Allbutt, H. A., 155
Allen, Grant, 74
Allman, G. J., 70
Anesthesia, 67, 77–78
Animal Magnetism (Binet and Féré), 130
Anthropogenie (Haeckel), 47
Anthropometry, 96
Anti-Semitism, 221, 223–37

Anus and anality, 142, 152; and excrement, 127, 131; Freud's theories regarding, 169, 174, 179–82, 190
Anxiety, 3–4, 7
Apollon (French Hercules), 103–04
Apollonian spirit, 88–89
Art, the body in, 21–33, 204, 254, 264–66; display of pubic hair in, 22–25, 27; influence of photography and motion pictures, 31–32; as sublimated sexuality, 75; asceticism as obstacle to, 88–93; modern, 191; and racism in Germany, 223–26
Art and Race (Schultze-Naumberg), 223–24
Art des parfums (Thoré), 52
Asceticism, 59, 78; in philosophy of Nietzsche, 88–93; as obstacle to art, 89–93
Auto-erotism, 197
Avenarius, Ferdinand, 223

Bachofen, Johann, 98
Balázs, Béla, 32

Balzac, Honoré de, 125
Bartels, Adolf, 223, 230
Bathers (Cézanne), 29
Bathers (Courbet), 23–24
Bathing, 35–36
Baudelaire, Pierre Charles, xii, 12, 26–27, 47, 52, 81–82, 252, 265
Bear, K. E. von, 63
Beardsley, Aubrey, 25, 280
Beauty, 92, 163; appreciation of, x–xi, 74–75; vs. ugly in art, 25–28, 264; Nazi concept of, 223–26, 228–29, 231; modern striving for, 256
Becker, Ernst, 65
Beckmann, Max, 225
Beerbohm, Max, 12
Before Sunrise (Hauptmann), 116
Behaviorism, 241–42
Behn, Aphra, 1
Being and Nothingness (Sartre), 243–47
Being and Time (Heidegger), 240
Beneden, Edouard van, 126
Benkert, Dr., 147
Berg, Leo, 97
Bergson, Henri, 239, 290–91
Bernard, Claude, 71, 83, 128, 252, 270
Bernard, Léopold, 53
Bestiality, 87, 138
Beyond Good and Evil (Nietzsche), 89, 92
Bicycling, 101, 121, 254
Binet, Alfred, 49, 125, 130, 141
Biology, basis of human existence, 60, 62–65; materialist views, 69–70; of women, 98, 162; and human sexuality, 126–27, 136, 162
Birth control, 98, 154–57, 160, 235; *See also* Contraception
Birth of Tragedy, The (Nietzsche), 88–89
Bisexuality, 135, 149–50, 187, 281
Blackwell, Elizabeth, 1–2
Blaikie, William, 103
Bleyl, Fritz, 30
Blind Mother (Schiele), 31

Bloch, Iwan, 47–49, 125, 131, 133–34, 141, 156, 165, 267, 280
Bloomers, 13, 251
Blue Angel, The (film), 32
Bodily fluids, 109, 112–13, 115
Body and Mind (Maudsley), 73
Body and mind dualism; *See* Mind-body dualism
Body odor, 45–55, 78, 87, 110, 151, 166, 251; thought to be cause of disease, 38, 46, 51; role in sexual life, 46–48; bad breath as grounds for divorce, 48; origin of the soul, 50–51; catalogues of, 51; expressed in literature, 52–55
Bölsche, Wilhelm, 125–28, 267
Bosom ring, 97
Boucher, François, 21
Bourgeoisie, 12, 58; morality of Victorian, 5–9
Bowdler, John, 5
Brain functions, 75–76
Brain size, 96
Brassiere, 17
Breasts, 13, 14, 17, 97, 179, 262; and Nazi concept of beauty, 224, 225, 228–29
Breath, bad, as grounds for divorce, 48
Brecker, Arno, 225, 226
Brentano, Franz, 239
Brieux, Eugene, 116
Broch, Hermann, 17–18
Brown, Horatio, 280
Brown, Isaac Baker, 101–02
Brown, Norman O., 189
Brown-Séquard, Charles, 128
Browning, Robert and Elizabeth, 3, 6
Brummel, George (Beau), 12
Buchanan, James, xii
Buchanan, Robert, 82
Büchner, Ludwig, 69
Budd, William, 46
Burial at Ornans (Courbet), 23
Burrows, Tom, 103

INDEX 295

Bustle, 11
Butler, Josephine, 44
Butler, Samuel, 114
Byron, Lord, 260

Camille (Dumas), 39
Camus, Albert, x
Cannibalism, 202
Capitalism, 144; and Victorian morality, 5, 7–8; Marx's theories regarding, 58, 60–62; sexuality in capitalist society, 235–36
Caplin, Roxey, 13
Caress, Sartre on the, 248–50
Carlyle, Thomas, 12, 18–19
Carpenter, Edward, 150, 280
Carpenter, William, 271
Castration, 102–03, 178, 184–85; of bodies of dead soldiers, 210, 286
Catharsis, 172
Cathexis, 172
Censorship, 133, 207–09, 214, 217
Cézanne, Paul, 29
Chadwick, Edwin, 36, 38
Chambard, Ernest, 130
Character, of women, 95–103; of men, 103–05; bodily determinants of, 179–87, 228
Charcot, Jean, 130, 141, 147, 171, 190, 253
Chastity, 5–7, 99, 101, 140, 195–96, 235–36; *See also* Virginity
Childbirth, 83, 184; Victorian attitudes regarding the role of women, 3–4, 96, 97, 100, 109–10, 153; importance of midwives, 37–38; God's will, 78; sex education and, 160–61, 164; in Nazi Germany, 233, 236–37
Children, 253; relationship with parents in Victorian family, 109–10, 123–24; hereditary transmissions and degeneration, 112–17; Schreber case, 117–19; child sexuality (masturbation), 119–22; sex education for, 157–62, 165; Freud's theories regarding sexuality in, 165, 179–85, 242; Hitler groups, 233, 236–37
Cholera, 38, 41, 46
Christian Science, 75, 252
Christianity, 8, 42, 57–58, 78, 140, 214
Circumcision, 120
Circus strong men, 103–04
Civilization and history, 187–90
Civilization and Its Discontents (Freud), 8, 188, 189
Clark, Kenneth, 28
Class distinction, 137
Class prejudice, 110
Class structure, 5–6, 92–93, 235–36
Cleaning the Temple of Art (Willrich), 225
Cleanliness, 34–36, 46
Clitoris, 99, 163, 166–67; surgical removal (clitoridectomy), 101–03, 120; Freud's theories regarding, 169, 174, 180, 183–86
Cloquet, Hippolyte, 46
Clothing, 222, 223, 251, 254; and sexual morality, 2–3; history since 19th century, 10–20; corsets, 2–3, 10, 13–16; changes in military uniform, 17; fashion in World War I era, 199–200
Coitus interruptus, 155–56
Cominos, Peter T., 7
Compayré, Gabriel, 115
Comstock, Anthony, 208
Conception, 109, 113
Concerning the Spiritual in Art (Kandinsky), 30
Condillac, Etienne de, 68
Condoms, 155, 251
Confessions (Rousseau), 217
Connolly, John L., Jr., 22
Contagion, fear of, 38–39; Contagious Disease Acts, 44
Contraception, 3, 7, 98, 133, 134, 153, 195, 202, 208; development of techniques of, 154–57
Conversion (psychology), 171–76, 190
Correspondances (Baudelaire), 52

Corset, 2–3, 10, 13–17, 97, 121–22, 205, 223, 263; call for reform of, 13; health problems of, 14–16; demise of, 16–17
Corydon (Gide), 151
Cosmetics, 200
Courbet, Gustave, 23–25, 264
Courtship, 4, 110–11
Creative discipline, 88–93
Creditors, The (Strindberg), 114
Crinoline, 11, 262
Critique of Pure Reason (Kant), 239
Crystal Palace Exhibition (1851), 35–36
Cubism (art), 29
Culture of the Female Body as a Foundation for Women's Clothing, The (Schultze-Naumberg), 14–15

Dancing, 93, 254
Dandyism, 12, 262–63
Darré, Gunther and Walther, 223, 233
Darwin, Charles, 46, 47, 56, 62–66, 70, 76, 226–28, 252
Das Kapital (Marx), 60–62
Das Weib (Ploss), 131
Dashwood, Sir Francis, 260
Davy, Sir Humphrey, 77
Death, ix, 39–41
Decadent movement, 55
Degas, Edgar, 27, 32
Degeneration, hereditary, 114–17
Déjà vu experiences, 178–79
Delacroix, Eugene, 23, 31
Delisle, Françoise, 132
Der Ablauf des Lebens (Fliess), 114
Der Menschenkenner, 228
Descartes, René, 68, 70, 241, 250
Descent of Man, The (Darwin), 46, 62–66
Desire, Sartre on, 248–49
Dessoir, Max, 129
Diaphragm, 155
Diderot, Denis, 68
Die Bläue Reiter, 30

Die Brücke, 30
Dietrich, Marlene, 32
Dinter, Arthur, 230
Dionysian spirit, 88–89
Diphtheria, 38, 41
Disease, 63, 266; contagion and human relations in Victorian times, 37–44; thought caused by odors, 38, 46, 51; love as subject of, 72; mind and, 73; Christian Science view of, 75; from masturbation, 101; Victorian attitudes toward germs, 112–17; homosexuality as, 147; Sartre's views on, 245
Divorce, 124, 133, 196, 204; bad breath as grounds for, 48
Doll's House, A (Ibsen), 205
Douching, 155
Douglas, Alfred, 145
Dreams, Freud's theories regarding, 170, 171, 177–79, 190; Scherner's symbols, 284
Dresses, 10, 16, 17
Drysdale, Charles, 132
Drysdale, George, 98
Du Bois-Reymond, René, 69
Dubos, René, 35, 37
Duncan, Isadora, 254
Dying (Schnitzler), 39

Earth (Zola), 208
Eberhard, E. F. W., 193
Economic and Philosophic Manuscripts (Marx), 58
Eddy, Mary Baker, 75
Ego, x, 57, 103, 202, 216–20
Ehrenfels, Christian von, 165
Ehrlich, Paul, 41, 231
Ellenberger, Henri, 281
Ellis, Havelock, 48, 71, 101, 125, 131–33, 136, 141, 149, 267, 271, 280
Embrace, The (Schiele), 30
Emotion, 185; artistic portrayal, 29–31; Darwin's theories regarding, 63, 76; and health, 73
Employment of women, 16–17

Engels, Friedrich, 57
England, Victorian sexual morality in, 4–5; clothing in, 16–17; development of public health facilities, 35–36; physiology of Victorian family, 109–24; scientific study of sex in, 125, 133
Epilepsy, 101
Erdmann, Karl, 263
Erikson, Erik H., 277–78
Eros and Civilization (Marcuse), 189
Eros in Barbed Wire (Henel), 197
Erotogenic (erogenous) zones, 128–31, 180–87, 190, 206, 253
Essence of Christianity, The (Feuerbach), 57
Etty, William, 265
Eulenberg, Alfred, 129
Evolution, 56, 62, 70, 74, 76, 96
Evolution of Sex, The (Geddes), 162
Excremental organs, 169, 174, 181–82
Existential thought, 238–50
Expression of Emotions in Man and Animals, The (Darwin), 76, 227
Expressionism (art), 25, 28–32, 223–26
Eyes, as index to personality, 76–77

Family, The (Schiele), 30
Family, Victorian, 109–24
Fantasy, 197
Faraday, Michael, 77
Fashion (style), 10–20
Fear, sexual, 196, 202
Feces, 181–82
Fechner, Gustav, 74
Fellatio, 152, 175
Female Sexuality (Freud), 185
Féré, Charles, 49, 125, 130, 267
Ferenczi, Sandor, 231
Fertility and menstruation, 155
Fetishism, 49, 140–41, 279
Feuer, Lewis, 269
Feuerbach, Ludwig, 57–60, 269; Marx's theses on, 59–60

Fischer-Dückelmann, Anna, 163–64
Flapper era, 17
Flaubert, Gustave, 207–08, 265
Flechsig, Paul, 102
Fleshly School of Poetry, The (Buchanan), 82
Fliess, Wilhelm, 47, 114, 125, 134, 142, 177, 187, 281
Flowers of Evil (Baudelaire), 26–27, 52, 82
Flugel, J. C., 19–20
Fol, Hermann, 126
Forel, August, 136, 164–65
Foreplay, 137, 167, 206
Fount of Life communities, 233
Fowler, F. N., 76
Fowler, O. S., 122–23
Fox, The (Lawrence), 209
Fragonard, Jean-Honoré, 21, 265
France, sexual morality in, 2, 4–5; bathing in, 35; scientific study of sex in, 125, 133; racial thinking in, 227
Freedom, fear of, 236, 245–50
Freud, Sigmund, 8–9, 30, 47, 59, 72, 75, 125, 134, 141, 154, 156, 166, 207, 214, 231, 242, 254, 281, 283–85, 287; on sense of smell, 49; Schreber case as material on paranoia, 117–19; theory of erotogenic zones, 128–31; on masturbation, 133; penis envy theory, 136; on sexual knowledge, 165; theory of anatomy of human sexuality, 169–90, 197, 242, 250; conversion symptoms, 171–76; corporeal determinants of dream symbols, 177–79, 215; bodily determinants of character, 179–87; civilization and the body, 187–90; laughter as release of energy, 212; Reich and, 234–35
Frick, Wilhelm, 224
Frigidity, 99, 105, 274
Fruits of Philosophy (Knowlton), 155
Fuchs, Eduard, 19, 104
Function of the Orgasm, The (Reich), 236

Gall, Franz Joseph, 75–76, 228
Gallo, Emanuele, 193
Galopin, Auguste, 46–47
Galton, Francis, 122
Garnier, Pierre, 102, 105–06, 162
Gauguin, Paul, 224
Gay Science, The (Nietzsche), 92
Geddes, Patrick, 96, 162
Genealogy of Morals, The (Nietzsche), 88, 90, 92
Genitals, 47, 122, 163, 213, 235, 253; displayed in art, 23, 30; and size of nose, 48; size of, 105, 106; and scientific study of sex, 129–30, 135; genital tissue, 141–42; genital kiss, 167, 206; Freud's theories regarding, 169, 174, 177–80, 183–87
Gerhard, William, 35–36
Géricault, Théodore, 23
German Ideology (Marx), 56, 60
Germany, Nazi culture of body politics, 15, 232–37; social function of clothes in, 18; expressionists in, 29–30, 32; and philosophy of Marx, 57–58, 60; scientific study of sex in, 125, 133–34; homosexuality in, 133, 149; moral revolution in, 193–95, 200–05; racism in, 203, 221–34, 236–37, 255; art criticism in, 223–26; sexuality in, 224–34, 236–37
Germinal (Zola), 84, 87
Ghosts (Ibsen), 115–16
Gide, André, 81, 150–52, 242
Giessler, Carl, 51
Giorgione, 24
Girls, what they knew in Victorian times, 157–62
Gobineau, Joseph A., 227
Goebbels, Joseph, 224, 230
Goethe, Johann Wolfgang von, 127
Gogh, Vincent van, 23, 25, 224
Goncourt, Edmond and Jules, 98
Gonorrhea, 41, 111, 156–57, 204
Goya, Francisco, 22
Grabinski, Bruno, 193
Grand Odalisque (Ingres), 22

Gratiolet, Louis, 227
Great Stench (1858), 46
Guilt, 84–85, 91, 93, 189, 201
Gunther, Hans F. K., 223, 224, 230

Haeckel, Ernst, 47
Haemmerling, Conrad, 144
Hair, 11, 17
Hall, G. Stanley, 121, 154
Haller, John S. and Robin M., 7, 104–05, 107, 273
Hamilton, Mary Agnes, 195
Hanna, Thomas, 269
Hartmann, Eduard von, 113
Hartmann, Geoffrey H., 265
Hauptmann, Gerhart, 116
Head-holder, 117–18
Health, 14–16; *See also* Physical culture movement
Health Culture and the Sanitary Woolen System (Jaeger), 51
Hearing, 74
Heckel, Erich, 30, 225, 265
Hedonism, 89, 92, 93
Hegar, Alfred, 102, 129
Hegel, Georg Wilhelm, 227
Heidegger, Martin, 240, 242, 248, 291
Helmholtz, Hermann, 74
Helvétius, Claude, 68
Henel, Hans, 197
Heredity, 83, 126, 149, 253; Victorian beliefs regarding, 112–17, 122; Krafft-Ebing's views on, 140–41, 147; Nazi views on, 230, 232
Hermann, Emanuel, 13–14
Hertwig, Otto, 126
Himmler, Heinrich, 232, 233
Hinton, James, 71
Hirschfeld, Magnus, 125, 133–34, 149–50, 193, 196–97, 200, 285
Hirth, Georges, 271
Hitler, Adolf, 222–25, 229, 255; phrenology, 228; anti-Semitism of, 231–33, 236–37
Holbach, Paul, 68
Holmes, Oliver Wendell, 37

INDEX

Homosexuality, 36, 129, 131–33, 137–38, 193, 194, 197, 253, 255, 281; in poetry of Whitman, 81; scientific study of, 133, 135; sexual pathology and, 139–52; causes of, 146–49; Freud's theories regarding, 185
Hooker, William, 72–73
Hormones, 128, 149
How to Get Strong and Stay So (Blaikie), 103
Hughes, H. Stuart, 152, 285
Hugo, Victor, 25, 26
Human nature, Marx's theory of, 57–62; Darwin's theory of, 62, 64–65; materialist theory of, 67–69, 71, 74–77; reshaped by the machine, 79
Humor in sexuality, 209, 212–13, 217–20, 255
Hunter, John, 41
Husserl, Edmund, 240
Huxley, Thomas, 70, 72, 252
Huysmans, Joris Karl, 23, 47, 52–55
Hyperaesthetic zones, 142
Hypnotism, 130–31, 171
Hysteria, 101, 130, 135, 157; drastic treatment for women with, 102; Freud's theories on, 171–76
Hysterical paralysis, 171–72, 175
Hysterogenous zones, 130, 190, 253

Ibsen, Henrik, 115–16, 193, 204, 205
Ideal Marriage (Van de Velde), 166–68
Immoralist, The (Gide), 151
Impotence, male, 132, 137, 156, 175; masculinity and sex role, 95, 100, 103–08, 252, 257; psychical, 186, 188, 254; brought on by war, 196–98, 202
Incest, 36, 187
Infanticide, 201
Infibulation, 120
Infidelity, 4, 140, 201, 203–05, 232
Ingres, Jean-Auguste, 22–24
Instinct, 140, 187–89

Institut für Sexualwissenschaft, 134
Intercourse, 48, 166–67, 186, 201, 203; women's role in, 99, 105–07; scientific study of, 133; in marriage, 140; dream symbols for, 178; Reich's theories on, 234
Interpretation of Dreams (Freud), 177–79

Jack the Ripper, 142–43
Jaeger, Gustav, 50–51
Jahn, Friedrich, 222
Jahrbuch für Sexuelle Zwischenstufen, 149
James, William, 38
Javanese women, 15
Jesus Christ, 213–14
Jews, in writings of Nietzsche, 92; sexuality of, 136, 203; victims of racism in Germany, 221, 223–37; Nazi laws against, 232
Joyce, James, 208

Kama Sutra, 168
Kandinsky, Wassily, 30
Kant, Immanuel, 46, 89, 239
Kerval, P., 130
Kinsey, Alfred, 126, 137–38
Kirby, Georgiana, 113
Kirchner, Ernst, 30, 225, 265
Kisch, E. Heinrich, 203–04
Kissing, 167, 181, 183
Kit's Woman (Mrs. Ellis), 132
Klimt, Gustav, 30, 265–66
Knowledge, 68–69, 241
Knowlton, Charles, 155
Koch, Robert, 38
Krafft-Ebing, Richard von, 49, 125, 126, 130, 131, 135, 139–47, 204, 253
Krauss, Friedrich, 231
Kretchmer, Ernst, 228
Kupfer, Amandus, 228

La Bête (Zola), 84
La Boheme (Puccini), 39
La Confession de Claude (Zola), 83–84
La Curée (Zola), 53
La Faute de l'Abbé Mouret (Zola), 53, 84–85

La Parfum de la femme (Galopin), 46
La Source (Ingres), 24
La Terre (Zola), 87
Labor, 252; Marx's theories regarding, 56, 60; alienation of, 60–62; replaced by machines, 79, 103
Lady Chatterley's Lover (Lawrence), 168, 208–14, 255
Lallemand, Claude-François, 104
Lamarck, Jean Baptiste, 113, 126
Lamettrie, Julien de, 68
L'Amour (Michelet), 97
Land and People (Riehl), 227
Lapouge, Vacher de, 35, 227
L'Assommoir (Zola), 53
Lawrence, D. H., 168, 207–15, 255, 287
Le Gousset (Huysmans), 54
Le Ventre de Paris (Zola), 53
League of Struggle for German Culture, 223, 224
Leaves of Grass (Whitman), 80
Leçons sur les Maladies du Système Nerveux (Charcot), 130
Lees, Edith, 132
Legs, 2, 32, 82, 205, 213
Les Avaries (Brieux), 116
Lesbianism, 209
Libido, 172, 202
Liebenfels, Lanz von, 229–31
Life Against Death (Brown), 189
Life in Nature (Hinton), 71
Lissmann, P., 196
Lister, Joseph, 38
Literature, sexuality in, 82–88; obscenity charges and censorship, 207–09, 214, 217; sexuality in works of Lawrence and Miller, 207–20
Little Olympia (Pouault), 28
Locke, John, 67
Lombroso, Cesare, 97, 98
Long, Crawford, 77
Lotze, Hermann, 19
Love, 62, 127, 139, 140, 181, 182, 185, 188, 195, 200, 230, 270; masochism as, 8; associated with disease, 39–40; smell as essence of, 46–47, 50; deification of, 57; nature of, 72; in Victorian marriage, 122–23; homosexual, 148–50; Sartre on, 246–50
Love and Its Hidden History (St. Léon), 72
Love and Parentage Applied to the Improvement of Offspring (Fowler), 122–23
Love Life in Nature (Bölsche), 126–28
Lower Depths, The (Gorki), 39
Lukacs, Georg, 273
Lust, 89

Macfadden, Bernarr, 104
Machines, 79, 103
Madame Bovary (Flaubert), 207–08
Madeleine Férat (Zola), 83
Magendie, François, 274
Magic Mountain, The (Mann), 39–41
Magnan, Velantin, 141, 147
Male Continence (Noyes), 156
Malthus, Thomas, 154
Man, sexuality and role, 4, 6–7, 103–05, 109–12, 120–21, 140, 143–44, 166, 200–06; clothing, 11–12, 16; nude in art, 23, 30; Darwin's views on, 64; impotence, 95, 100, 103–08, 186, 188, 196–98, 202, 252, 254, 257; virility and physical exercise, 103–04, 107; masturbation, 110; in war, 194–99, 202
Man a Machine (Lamettrie), 68
Man Who Died, The (Lawrence), 214
Manet, Edouard, 24, 25, 29, 31
Mann, Thomas, 39–41, 116–17
Mantegazza, Paul, 48
Marcel, Gabriel, 240, 242, 250, 290
Marcus, Steven, 6
Marcuse, Herbert, 189
Marholm, Laura, 98
Marquis de Sade and His Time, The (Bloch), 133–34
Marriage, 58, 140, 235, 252; and Victorian morality, 3–5, 7, 8; Victorian marriage manuals, 3, 110–13, 122–23, 155–60; passive sexual

role of women, 99–100; physiology of Victorian, 109–12, 123–24; sex life in, 140, 162, 164; *Ideal Marriage*, 166–68; postponement as moral restraint, 154; moral revolution in, 193, 196, 204–05; in Nazi Germany, 232–33
Marx, Karl, 56–62, 65–66, 227, 234, 252, 269
Masochism, 87, 143–44, 180; Sartre on, 246–47, 249–50
Mass Psychology of Fascism, The (Reich), 236
Masturbation, 3, 105, 106, 110, 132, 138, 142, 147, 152, 200, 204, 216, 255; by women, 100–02, 158, 160; diseases from, 101; child sexuality, 119–21, 156; scientific studies of, 133; Freud's theories regarding, 170, 178, 183–85
Materialism, 57, 59–60, 83; and mind-body problem, 67–79
Matter and Memory (Bergson), 239
Maudsley, Henry, 49, 73, 98, 252
Mayreder, Rosa, 98–99
McLuhan, Marshall, 32, 79, 270
Mechanist views, 69–71
Medicine, 1–2; effects on human relations, 34, 37–44; Christian Science rejection of, 75; technology and progress, 195
Mein Kampf (Hitler), 231
Meisel-Hess, Grete, 202
Memory, 49–50, 73, 118
Mencken, H. L., 9
Mensinga pessary, 155
Menstruation, 4, 47, 110, 200; key to female temperament, 97–98; sex during, 111, 156–57; and fertility, 155
Mental illness, 188–89
Mercury, 41, 43, 105, 128
Merleau-Ponty, Maurice, 238–39, 242, 250, 254, 269, 290
Metabolic rates, 96
Metaphysical Journal (Marcel), 240

Methodism, 8
Michael (Goebbels), 230
Michelot, Jules, 97–98
Michels, Robert, 111
Middle Ages, sexuality in, 6
Midwives, 37
Miller, Henry, 194, 207, 214–20, 255, 287
Millet, Jean-François, 23
Millett, Kate, 287
Milton, J. L., 104
Military uniform, 17–18, 200
Mimic and Physiognomy (Piderit), 227
Mind-body dualism, 68–69, 72–79, 98, 162; Darwin's theories regarding, 64, 65; Freud's theories regarding, 170–90; Freud's dream theories, 177–79; Freud's theories on body determinants of character, 179–87; existential philosophy, 238–50
Miscegenation, 229–32
Möbius, Paul, 96–98, 134, 135
Modern Times (film), 61
Moers, Ellen, 262–63
Moll, Albert, 125, 129, 131, 133, 147
Monin, E., 266
Monogamy, 5, 8, 165, 186, 188, 230
Morality, sexual, 1–9, 24; in writing of Zola, 83–84; in writing of Nietzsche, 92–93; revolution in, 192–206; civilized repression of sex, 171, 188–89, 207–09, 235–36
Moreau, Paul, 147
Morel, B. A., 114–15
Morton, William, 77
Motion pictures, 32
Mueller-Lyer, F. C., 131
Müller, Fritz, 16
Mumford, Lewis, 7–8, 79, 205
Munch, Eduard, 30
Music, 88, 191
Mutterschutz, 134
Myth of the Twentieth Century, The (Rosenberg), 224

Naked Maja (Goya), 22
Nakedness; *See* Nudity
Nana (Zola), 53, 84–87
Napheys, George H., 99, 111, 156
Napoleon, 283
National Life and Character (Pearson), 227
Nausea (Sartre), 243–45
Nazism, 15, 134, 136, 137; use of phrenology to support theories, 77; racism of regime in Germany, 222–34, 236–37
Nerves, 196, 254
Neuroses, 156, 173–76, 187
New Olympia (Cézanne), 29
New York Society for the Suppression of Vice, 208
Nicolson, Harold, 5
Niederland, William G., 118
Nietzsche, Friedrich, 45, 48, 59, 88–93, 127, 189, 238, 271, 273
Nightingale, Florence, 46
Nihilism, 91–93
Nocturnal emissions, 3, 104–05, 132, 138, 156, 196
Nolde, Emil, 225
Nose, 77, 141–42
Noyes, John Humphrey, 155–57
Nudity, 3, 6, 51, 133, 193, 204, 225, 232, 249, 251, 254; in art, 21–33, 264–66; pubic hair in art, 22–25, 27, 265
Nuremberg Laws (1935), 232
Nymphomania, 101, 200

Obscenity, 196, 249
Obsession, 175–76
Odor; *See* Body odor
Oedipus complex, 186, 209
Olfactometer, 51
Olfactory ontology, 45–55, 267
Olympia (Manet), 24, 31
Olympic Games (1936), 222–23
On the Physiological Feeblemindedness of Women (Möbius), 96–97
Oral sex, 82, 167, 181

Orgasm, 7, 130, 137–38, 186, 209–10, 252; by women, 3, 95, 99, 100, 102, 261, 274; Freud's theories regarding, 173, 179–80; Reich's theories regarding, 234–35
Orgone, 234
Origin of Species (Darwin), 62–64
Orwell, George, 194
Osphrésiologie (Cloquet), 46
Ostara (Liebenfels), 229
Ovaries, removal (castration), 102
Owens, Jesse, 222–23

Pain, ix–x, 143, 195; psychology of, 67–68, 74; anesthesia, 77–78; and perversion, 129; Freud's theories regarding, 174–76; Sartre's theories regarding, 244–47, 249
Painter's Studio, The (Courbet), 23
Paralysis, organic and hysterical, 171–72
Paranoia, 118–19
Parent-child relationship, in Victorian family, 109–10, 123–24; hereditary transmission and degeneration, 112–17; Schreber case, 117–19; child sexuality, 119–22; Freud's theories regarding, 175, 183–86, 242; Oedipus complex, 186, 209
Pasteur, Louis, 38
Paulk, Emil Gustav, 202–03, 210
Paull, Hermann, 230
Pearson, Karl, 227
Penicillin, 43
Penis, 122, 136, 166–67, 286; Freud's theories regarding, 169, 173–75, 177–78, 180, 183–84; phallic symbols, 177–78; penis envy, 184, 185, 187, 287; in literature, 211–12, 214–16; measurement, 215–16
Perception, 241
Perfume, 52, 54
Personality, determined by head shape and face, 75–77; Freud's

theories regarding development of, 179–87; Nazi cult of, 221–34
Perspiration, 51, 54
Perversion, 129, 143–44, 179–83
Pessary, 155
Peters, Emil, 107
Petticoat, 205
Phallic symbols, 177–78
Phenomenologist thought, 240–44
Philosophy of the body, 238–50
Phobia, 175
Photography, 31–32
Phrenology, 75–77; used by Nazis to justify actions, 77, 227–28
Physical attraction, 256
Physical culture movement, 103–04, 107, 254, 255; Schreber case, 117–19; Nazi philosophy, 222, 223, 233
Physiognomy, 76, 227
Physiognomy and Expressive Movements (Gratiolet), 227
Physiological Feeblemindedness of Women, The (Möbius), 135
Physiology and Pathology of the Mind, The (Maudsley), 49
Physiology of Smells (Zwaardemaker), 51
Physique and Character (Kretchmer), 228
Picasso, Pablo, 29
Picnic on the Grass (Manet), 24
Piderit, Theodor, 227
Pitres, Albert, 130
Place, Francis, 155
Plato, 241
Pleasure, 3, 8; ethic of, 67–68, 74, 77–79
Ploss, Heinrich, 131
Poetry, 52; reality in, 25–27, 252; sexuality in, 80–82
Polygamy, 165
Poor, immorality of, 36
Pornography, 31, 82, 137, 193, 196, 208, 212
Post-Impressionism (art), 28–29

Pregnancy, 3, 109, 112–14; sex during, 113, 157, 276
Pre-marital sex, 137, 201
Premature ejaculation, 197, 198
Prisoners of war, 200, 201
Privacy, 36–37
Problem of Modern Ethics, A (Ulrich), 148
Promiscuity, 201
Propaganda, 201, 223
Prostitution, 134, 137, 150, 255; Victorian attitudes toward, 3, 4, 8, 36, 58, 110; effects of Contagious Disease Acts on, 44; Freud's theories regarding, 183; during World War I, 193, 197–99; in Germany, 200, 216–20, 232
Protestantism, 8, 120
Protoplasm, 70
Proust, Marcel, 50, 125, 242
Prudery, 2, 6, 48, 99, 208, 260
Psychic energy, 172–73
Psychoanalysis, 120, 165, 206, 242, 254; Freud's theories of, 171–90
Psychology From an Empirical Standpoint (Brentano), 239
Psychopathia Sexualis (Krafft-Ebing), 126, 139, 143–44, 253
Puberty, 135, 142, 183–85
Pubic hair, 178, 212, 225; in art, 22–25, 27, 265
Public executions, 78
Public health, 34–37, 46, 121, 251
Pudor, Heinrich, 51
Puerperal fever, 37–38
Puritanism, 9

Queensberry, Marquis of, 145
Quennell, Peter, 11
Quinlan, Maurice, 6

Race and Style (Gunther), 223
Racial Beauty of Women, The (Stratz), 229
Racism, 114; racial or national smells, 49, 50; in scientific studies, 136,

137; in Germany, 165, 203, 221–34, 236–37, 255; sexual, 167
Raft of the "Medusa," The (Géricault), 23
Ranke, Leopold von, 191
Rape, 202
Rasso Trio, 104
Rau, Hans, 129, 130
Realism (art), 23–25
Rebbing, Seved, 162
Reich, Wilhelm, 5, 231, 234–37
Religion, 53, 59, 60, 195; and Victorian sexual morality, 3, 5, 8; and Darwin's theory of evolution, 62; controversy over anesthesia, 78; and sexual feeling, 140; God in sex education, 159–60
Remarque, E. M., 195
Rembrandt, 21
Remembrance of Things Past (Proust), 50
Renoir, Jean, 29
Report on the Sanitary Conditions of the Labouring Population of Great Britain (Chadwick), 36
Reproduction, scientific study of, 126–27, 129, 135–36, 252
Revolution, 59–60; of morality, 192–206
Rhythm, 155
Ribot, Théodule, 49–50
Richter, Johann Paul, 130
Ricord, Philippe, 41
Riefenstahl, Leni, 223
Riehl, Wilhelm, 227
Rilke, R. M., 265
Rochard, Jules, 78
Rodin, Auguste, 27, 28
Rohleder, Hermann, 133
Roles of men and women, sexual, 3–7, 16–17, 95–108, 110–12, 140–43, 209–20, 254–55, 257
Rosenberg, Alfred, 223, 224, 230
Rosenkranz, Karl, 25–26
Rossetti, Dante Gabriel, xii, 82, 252
Rouault, Georges, 27–28, 32

Rousseau, Jean Jacques, 125, 217
Rousseau, Théodore, 68
Rubens, Peter Paul, 21
Rugoff, Milton, 274

Sacher-Masoch, Leopold von, 143
Sachs, Hans, 231
Sade, Marquis de, 68, 82, 143
Sadism, 134, 143–44, 180, 196, 201–02; female, 193, 202; Sartre on, 246–50
St. Léon, Count de, 72
Salvarsan, 41
Sanger, Margaret, 132
Sanitary conditions, 34–37
Santayana, George, 74–75, 271
Sargent, Dudley Allen, 103
Sartor Resartus (Carlyle), 18
Sartre, Jean-Paul, 30, 239, 242–50, 256, 290
Schaudinn, Fritz, 41, 231
Scheffler, Karl, 193–94
Scherner, Karl Albert, 284
Schiele, Egon, 25, 30–31
Schiller, J. Friedrich von, 50, 139
Schmidt-Rottluff, Karl, 30, 225
Scholz, Friedrich, 129–30
Schopenhauer, Arthur, 68–69, 270
Schreber, Daniel Gottlob and Daniel Paul, 117–19, 176
Schreiner, Olive, 132
Schultze-Naumberg, Paul, 14–15, 223–26, 230
Scientific study of sex, 124–38
Scott, Sir Walter, 1
Self-control, 196
Self-denial, 58–59; philosophy of Nietzsche, 88–93
Self-destruction, 189
Semen, 156, 229–30; magical power attributed to, 109, 113
Semmelweiss, Ignaz, 37
Sensations of Tone (Helmholtz), 74
Senses, 45–55, 61, 89–90, 241
Sewer systems, development of, 35–36, 251

Sex and Character (Weininger), 134–35
Sex crimes, 142–43, 201–02
Sex education, 133, 134, 153–57, 208; development of contraceptives as aid to, 154–57; marriage manuals, 155–60; for Victorian girls, 157–62; positive, 162–68
Sexual Behavior in the Human Male (Kinsey), 137
Sexual Infidelity of Women, The (Kisch), 203–04
Sexual Instinct, The (Féré), 131
Sexual Inversion (Ellis), 149
Sexual Life of Our Time, The (Bloch), 134, 165
Sexual Politics (Millett), 287
Sexual Question, The (Forel), 136, 164–65
Sexualethik (Ehrenfels), 165
Sexuality, Victorian morality, 1–9, 24, 37, 41–45, 109–24; sensuality and eroticism in art, 21, 24, 28–30; generated by crowding, 36–37; of love and death, 40; role of olfactory system in, 46–49, 51; alienation, 56–62, 269; Darwin's theory of selection, 62, 64–65; materialist view of, 68–69; mental processes and, 73; influence on artistic creation, 75; in poetry, 80–82; in Zola's writing, 82–88, 94; influence on temperament, 83; roles of men and women, 95–108, 110–12, 140–43, 209–20, 254–55, 257; of Victorian children, 119–22; scientific study of, 125–38; study of sexual customs, 131; pathology and homosexuality, 139–52; development of contraceptives, 154–57; Freud's theories regarding, 169–90; civilization and, 187–90; transformation of morality, 192–206, 251–57; as cause of war, 193, 194; front-line bordellos, 197–99; in writing of Lawrence and Miller, 207–20; Nazi concept of, 224–34, 236–37; Reich's theories regarding, 234–37; Sartre's views regarding, 246–50
Sexuelle Neuropathologie (Eulenberg), 129
Shame, 6, 164, 201, 246
Shaw, George Bernard, 12
She Who Was Once the Helmet-Maker's Beautiful Wife (Rodin), 28
Shoe fetishism, 49, 144
Simpson, Sir James, 77–78
Sin, 8, 21, 111, 139
Sins Against the Blood (Dinter), 230
Sledge, 107
Sleepwalkers, The (Broch), 17–18
Slums, 36
Smell, sense of, 46–55, 78, 87, 110, 129, 151, 166, 195, 251, 267; role in sexual life, 46–49; in literature, 52–55; fetishes involving, 141–42; Freud's theories regarding, 176
Smith, Southwood, 38
Snow, John, 38
Snuff, 261
Sodomy, 197, 255
Somatic compliance, 172, 175–76
Sons and Lovers (Lawrence), 209
Soul, 50–51, 81
Spencer, Herbert, 96, 129, 226–27
Sperm, 127–28, 132
Spermatorrhea, 3, 104–05, 132, 156
Spice boxes, 54
Spitz, René, 120–21
Spokesman, body as, 119, 277–78
Sponges, contraceptive, 155
Spring's Awakening (Wedekind), 160–61
Spurzheim, Johann, 76, 228
Stall, Sylvanus, 107, 111, 157
Steinach, Eugen, 128, 231
Stekel, Wilhelm, 200, 231
Stendahl, 125
Sterilization, 230, 232
Stöcker, Helene, 196
Stoll, Otto, 131
Story, Alfred T., 77

Stratz, Christian H., 15–16, 225, 228–29, 261
Streicher, Julius, 230
Strength Through Joy organizations, 233
Strindberg, Johan August, 114, 135, 193
Studies in Hysteria (Freud), 173
Studies in the Psychology of Sex (Ellis), 48, 131
Study of a Nude Male (Géricault), 23
Stylish Beauty (Schultze-Naumberg), 225
Sucking impulse, 181
Suicide, 107, 201
Swift, Jonathan, 213
Swoboda, Hermann, 114, 281
Symonds, John Addington, 148–49, 280
Syphilis, 41–44, 115–16, 128, 201, 204; treatment for, 43–44; charge of Jewish origin, 231, 289
Syphilophobia, 42

Tarnowsky, Benjamin, 147
Taste, sense of, 45, 47, 50
Temperament, 83, 97–98
Testicles, 128
Teufelsdrockh, Professor, 18
Thérèse Raquin (Zola), 83
Thermes, Godefroy, 102
Theses on Feuerbach (Marx), 59–60
Thompson, E. P., 8
Thoré, Théophile, 52
Three Essays on the Theory of Sexuality (Freud), 131, 165, 179–83
Thuillier, Guy, 35
Thus Spake Zarathustra (Nietzsche), 91–93
Toilet training, 8, 119
Tolstoy, Leo, 193
Totem and Taboo (Freud), 131, 189
Toulouse-Lautrec, Henri-Marie, 25, 27–28, 32
Transmission of Life, The (Noyes), 156
Transvestism, 149–50, 197

Treatise on Degeneration (Morel), 114–15
Tristan (Mann), 116–17
Tropic of Cancer (Miller), 208–09, 214–20
Trousseau, Armand, 105
Tuberculosis, 38–41
Tuke, Daniel Hack, 271
Turkish Bath, The (Ingres), 22–23

Ugly, depicted in art, 25–28
Ulrichs, Karl, 147–48
Ulysses (Joyce), 208
Unconscious Memory (Butler), 114
Underwear, 50–51, 200
Une Page d'amour (Zola), 84
Uniform fetish, 17–18, 200
Upright-holder, 117
Urolagnia, 132
Uterus, 178

Vagina, 112, 218–19; Freud's theories regarding, 179–80, 183–86; sponges as contraceptives, 155
Vampirism, psychological, 202
Van de Velde, Theodore, 166–68, 206
Van Ussel, Jos, 6, 274
Veblen, Thorstein, 19
Venereal disease, 105, 137, 162, 165, 251; problem to Victorian man, 3, 110, 111, 115, 153, 156–57, 160; treatment for, 41–44; intercourse with virgin as cure for, 44, 267; during World War I, 193, 195, 197–99, 201, 204; Nazi views on, 231–32
Ventilation, 46
Venus in Furs (Sacher-Masoch), 143
Vertigo, 157
Victoria, Queen, 78
Victorian, sexual morality, 1–9, 24, 109; physiology of family, 109–24
Virchow, Rudolf, 98
Virginity, 112, 140; intercourse with virgin as cure for venereal disease, 44, 267

Virile Powers of Superb Manhood, The (Macfadden), 104
Vision, sense of, 32, 48
Vitalistic views, 69–71
Voyeurism, 36, 100, 144, 176, 246

Wagner, Richard, 88–89
War, 193, 196; effects of men and women, 194–205
Wassermann, August von, 41, 231
Water supplies, 34–36, 251
Watson, John, 183
Way of All Flesh, The (Butler), 114
Weber, Max, 59
Wedekind, Frank, 160–61, 193
Wegener, Hans, 157
Weininger, Otto, 113–14, 125, 134–36, 230, 231, 281
Weismann, August, 113, 126
Westermarck, Edward, 131
Westphal, Carl, 147
What a Young Girl Ought to Know (Wood-Allen and Stall), 157
Whitman, Walt, xii, 80–81, 93, 147–48, 252
Wife's Handbook, The (Mensinga), 155
Wilberforce, William, 5
Wilde, Friedrich, 155
Wilde, Oscar, 145–46, 150, 253, 280
Will power, 68–70
Willrich, Wolfgang, 225–26, 230
Wir Jungen Manner (Wegener), 157
Wollstonecraft, Mary, 7, 261
Woman of the Sea (Ibsen), 204
Woman Pulling Up Her Stockings (Toulouse-Lautrec), 27–28
Women, Victorian sexual morality of, 1–9, 109–12, 157–62; clothing, 2–3, 10–20; orgasm, 3, 95, 99, 100, 102, 261, 274; feminist movement, 6–7, 95, 103, 107–08, 192–96, 199, 202–06, 257, 287; of Java, 15; employment of, 16–17; nude in art, 21–27; bias of Contagious Disease Acts against, 44; Darwin's theories regarding, 64–65; God's will to endure pain, 78; sexuality and roles of, 95–103, 105–06, 110–12, 135, 140, 143–44, 157, 162–68, 192, 199–206, 209–20, 252–55, 257; evolutionary theory of, 96; masturbation by, 100–02, 158, 160; castration, 102–03; have greater need for love, 140; as homosexuals, 152; mother-daughter relations, 160–61; Victorian sex education for girls, 157–62; positive sex education for, 162–68; Freud's theories regarding female sexuality, 169, 183–87, 287; bisexual disposition, 185–86; frontline bordellos, 197–99; sexuality in writings of Lawrence and Miller, 209–20; German concept of beauty, 225–26, 228–30
Women in Love (Lawrence), 209
Women of Avignon (Picasso), 29
Wood-Allen, Mary, 157
Work ethic, 7–8
World War I, 153, 162, 189, 236; effects on morality, 16–20, 191–206, 229, 230, 236, 254–56

Yellow Book, The, 146, 280
Youth, 15; Hitler groups, 233, 236–37; *See also* Children

Zaner, Richard M., 290–91
Zeitschrift für Sexualwissenschaft, 134
Ziegler, Adolf, 225
Zola, Emile, 47, 52–53, 55, 82–88, 93–94, 115, 125, 208, 252
Zwaardemaker, H., 51